SCIENCE AND

Volume Two

SENSIBILITY

by *James R. Newman*

SIMON AND SCHUSTER · NEW YORK · 1961

Some of the material that appears in this book appeared in different form in the following publications; for permission to use this material, the author and the publishers are indebted to:

Scientific American, Inc., for the following articles: *Nature and the Greeks, Education in Antiquity, Science and Civilization in China, The Copernican Revolution, The Lisbon Earthquake, Freud, John Maynard Keynes, Francis Bacon, Cardano, the Gambling Scholar, Isaac Newton, The Marquis de Laplace, James Clerk Maxwell, Francis Galton, The Wright Brothers, Arthur S. Eddington, Srinivasa Ramanujan, Blaise Pascal, John Stuart Mill, Bertrand Russell, Ludwig Wittgenstein, The Age of Analysis, Schrödinger — Mind and Matter, Causality and Chance in Modern Physics, Determinism and Indeterminism, Reason and Chance in Scientific Discovery, A Million Years From Now, War or Peace, Atomic Harvest, The Wandering Albatross, The Foreseeable Future* and *Comets and Their Origins.* Copyright 1948, 1950, 1951, 1952, 1953, 1954, 1955, 1956, 1957 and 1958 by Scientific American, Inc. Copyright © 1958, 1959 and 1960 by Scientific American, Inc.

The Curtis Publishing Company for *Einstein,* © 1959 by The Curtis Publishing Company.

Alfred A. Knopf, Inc., for *William Kingdon Clifford,* reprinted from THE COMMON SENSE OF THE EXACT SCIENCES, by William Kingdon Clifford, by permission of Alfred A. Knopf, Inc. Copyright 1946 by Alfred A. Knopf, Inc.

The New Republic for the following articles: *The Scientific Attitude, The Commendablest Phrases, Blackstone for the Proletariat, Patterns of Panic, Physics and Politics, Mind Under Matter* and *Quest for the Mind.* Copyright 1947, 1948, 1949, 1950 by James R. Newman.

Permissions and copyrights for illustrative material will be found in the captions accompanying the individual illustrations.

LIBRARY OF CONGRESS CATALOG CARD NUMBER: 61-12869
MANUFACTURED IN THE UNITED STATES OF AMERICA
PRINTED BY
THE MURRAY PRINTING COMPANY, FORGE VILLAGE, MASS.
BOUND BY H. WOLFF, INC., NEW YORK

TO MY CHILDREN

Contents

Volume Two
PART 1

PART 2

PART 3

PART 4

List of Illustrations

PART
1

*Blaise Pascal. Portrait by Philippe de Champaigne.
(The Bettmann Archive)*

BLAISE PASCAL

D
ID RELIGION consume Blaise Pascal? Did it
lead him to immolate the intellect, to for-
sake reason for faith? He stands out for his achievements even
in the century of genius in which he lived, yet how much more,
one wonders, might he have accomplished had he not wasted
himself on theological sterilities and in religious quarrels? In
one of Nietzsche's eloquent polemics against Christianity it is
Pascal whom he holds up as a tragic victim of the ravages of
religion: "What is it that we combat in Christianity? That it
aims at destroying the strong, at breaking their spirit, at ex-
ploiting their moments of weariness and debility, at converting
their proud assurance into anxiety and conscience-trouble; that
it knows how to poison the noblest instincts and to infect them

3

with disease, until their strength, their will to power, turns inwards, against themselves — until the strong perish through their excessive self-contempt and self-immolation: that gruesome way of perishing, of which Pascal is the most famous example."

No doubt Pascal's struggle with himself, his attempt to reconcile religious faith with the spirit of geometry, and his final justification of religion on a nonintellectual basis, cost him dear. He was not a saint by nature. He had ferocious energy and a fiery temper (his sister Jacqueline tells us he had "*une humeur bouillante*" — a sisterly euphemism); he was a man of passion in his arguments, his friendships, his scientific exertions, his religious beliefs. He had a passion even for self-torture. But there are other facts that cast doubt on the opinion, perhaps rather widely held, that Pascal sacrificed his magnificent creative powers to his God. They are set forth in a lucid, insightful biography by an English clergyman, Ernest Mortimer.*

The subtitle of his book is "The Life and Work of a Realist," and he endows the word "realist," as a British reviewer has said, with its full breadth of meaning. Two of his major points should be mentioned at the outset. The first is that it is simply untrue that Pascal abjured science and society after his conversion in 1654; it is a neat picture, but the evidence does not fit it. Over the years his religious interests deepened, but until the very end of his life he clung to his "free-thinking friends" and retained his scientific curiosity. The second point is that Pascal held facts to be no less sacred than faith. He believed in sense data as a source of knowledge and truth; he recognized that the contradictory impressions 'of nature are no excuse for mystery-mongering: patient observation would explain them; he respected the power of the intellect and reason to enlarge the "gates of perception" upon the universe. In all

* Ernest Mortimer, *Blaise Pascal: The Life and Work of a Realist*, New York, 1959.

4

this he was a modern thinker. And he was no less modern, it can be argued, in believing that the emotions are another gate of perception. The heart has its reasons, which reason knows nothing of; modern psychology and psychoanalysis have said it, but not so well.

Pascal was born at Clermont in Auvergne on June 19, 1623. His father, Étienne Pascal, was a well-to-do high public official, a judge and tax commissioner, who was to rise even higher. Blaise's mother, who came of a family of prosperous merchants, died at the age of thirty, leaving him, aged three, with an elder sister Gilberte and a younger sister Jacqueline.

The boy was sickly and precocious. The nature of his illness is not fully known (a witch was said to have been partly responsible) but an anatomical anomaly discovered after his death was probably a contributing factor. Of the two apertures in a baby's skull, which should knit up in the early weeks of life, it was found that in Pascal's case one of them had never properly closed, and the other had clamped or overlapped so as to form a bony ridge. He suffered from headaches all his life. His precocity served him better. His family was bound close by natural affection and by bereavement; they loved him and recognized his exceptional powers. Blaise was "encouraged rather than forbidden to stare."

On the death of his wife Étienne Pascal resolved to be father, mother and tutor to his children. They had their governess, but he alone gave them their formal education. His instruction, says Mortimer, was remarkable for its paternal self-denial, its eccentricity, rigor and patience; and also for its fruits. He scorned "the sort of pedagogy which he had suffered himself"; he had his own theories of teaching and did not hesitate to practice them. His principal maxim, Gilberte tells us, "was to hold the child always above his work," not to rush him or overload him. Instruction was by easy conversation. The object was to quicken the child's interest, to let natural curiosity about language, about the things around him, guide the

5

direction and emphasis. It was time enough for a child to learn
Latin when he was twelve, when he could do it more easily.
"By good fortune," Pascal wrote to a friend in later years,
"for which I cannot be too grateful, I was taught on a peculiar
plan and with more than fatherly care."

Étienne had a bent for mathematics, and decided to make
that subject "the coping stone of his plan." Not until he was
sixteen, after he had learned Latin, Greek, history and geog-
raphy, was Blaise to be introduced to geometry. Meanwhile
books of mathematics were not allowed in the house, and the
subject was not discussed in the daily conversational instructions. But the boy had secretly made his own way. One day
when he was twelve his father came upon him surrounded with
diagrams, "so absorbed that he still thought himself alone."
He had been trying to work out for himself the principles of
geometry, calling straight lines "bars" and circles "rounds,"
and he was now trying to prove that the three angles of a triangle add up to two right angles. (Another version of the story
is that he had independently rediscovered the whole of Euclid
up to the 32nd proposition of the first book; this is implaus-
ible — though not impossible — if for no other reason than
that if he had read Euclid he would at least have known the
right names for lines and circles.) Étienne forgot his theoretical scruples and was overcome with pride and joy.

Admirable though it was, Blaise's education had its gaps.
History and literature were sketchily imparted, the natural
sciences even more sketchily. Fortunately Blaise was not over-
burdened with theology; for the time being, elementary religious teaching was enough. But he was soon to come out into
the world and to glimpse new horizons of learning: culture,
science — and theology.

When he was eight, the family had moved to Paris. Étienne
had rented a house in the Rue de Tisseranderie (which, it is
delightful to learn, was intersected by the Street of the Two
Doors, the Street of the Bad Boys, Cock Street and the Street

6

of the Devilish Wind) and had entered into the life of marvelous city. He now had few professional duties and cou. give up even more time to the education of his children. He was a sociable man, with a taste and talent for natural science, and he found friends with whom he could share his interests. Social groups in which the arts and sciences were discussed seriously "but outside academic settings," were a feature of the time. He was close to Guillaume Montory, the leading actor of the day, and to Charles Dalibray the poet; the fashionable circles of literature and art were open to him. More important was his membership in the Abbé Bourdelot's group, which included, among others, the aging but still lively mathematician and physicist, Père Mersenne ("a bold, capacious, untidy and extraordinarily vivacious mind"), who managed to reconcile his support both of Galileo and orthodox theology; the gifted but prosaic mathematician Personne de Roberval (who, it is told, having seen a performance of *Le Cid*, complained that he could not see what it proved); the brilliant young engineer, Pierre Petit, who with Étienne was later to reconstruct Torricelli's famous experiment; and Gérard Desargues, who invented projective geometry and introduced the method of perspective. Blaise was allowed, while still a child, to accompany his father to the séances of this group and even encouraged to utter his own ideas. He was much influenced by Desargues, not only in mathematics, but in a concern, which the older man felt deeply, for improving the education and lightening the labors of plain workmen.

Pascal's first contribution to mathematics, composed at the age of sixteen, was his celebrated theorem. He had read Desargues' writings and had recognized as no one else had the worth of Desargues' method. The little "Essay on Conics," which acknowledged his debt to Desargues, described a property common to all conic curves, that is invariant under projection and section: If any six points on the curve are joined by straight lines and the sides of the resulting hexagon are pro-

7

longed beyond the curve, the three pairs of opposite sides will intersect at three points, which lie on one straight line.

The theorem was widely praised, but not by the great Descartes. In reply to Mersenne's enthusiastic letter he wrote: "He seems to have copied Desargues. I cannot pretend to be interested in the work of a boy." It is true Desargues' treatise suggested the property, but Pascal "isolated and proved it." It was a beautiful piece of work in its own right, and not merely a feat of precocity.

And now he was no longer a child. The eight years in Paris had been "intensely formative." He was mastering Latin, beginning Greek, absorbing paternal disquisitions on philosophy at mealtimes. He had met leading scientific thinkers; he had had his first taste of *la vie mondaine*. His energy matched the fecundity of his intellect: it is said, for example, that he followed up his theorem on conics with some four hundred corollaries. Mathematics was unquestionably his great passion. He reserved it, to be sure, for his leisure moments; because it came easy to him, he may have regarded it as a self-indulgence. But as his sister Gilberte wrote, *"Il trouvait dans cette science la vérité qu'il avait si évidemment recherchée il en était si satisfait qu'il y mettait son esprit tout entier."* Like the young Bertrand Russell, he hoped to find in mathematics the solace and perfection of timeless truth.

In 1640, Étienne having through the favor of Richelieu been awarded an important but unpopular post as tax assessor for the city, the Pascals moved to Rouen. There they remained for seven years. For Blaise it was a period of "much hard work, a religious awakening and the beginnings of a profound spiritual dilemma."

Étienne's job was back-breaking: in addition to various administrative chores he had the task of reassessing the taxes of eighteen townships, which meant, among other things, copying endless columns of figures. "I never get to bed before two o'clock," he wrote Gilberte. The spectacle of his father drudg-

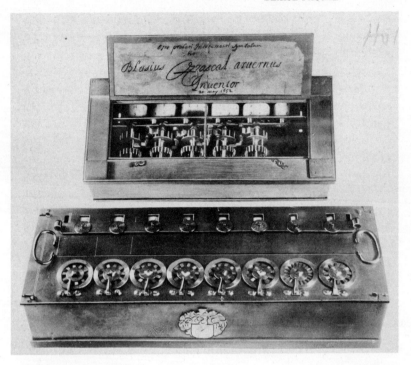

One of Pascal's machines à calculer. *(Musée de Conservatoire Nationale des Arts et Métiers, Paris)*

ing put an idea into Blaise's head. Why not make a machine to do the donkey calculations? During the first year at Rouen he conceived a mechanism that would perform all the operations of arithmetic; it would have a device for carrying digits from one column to another, and another for recording the result. With the help of local craftsmen he worked on the idea. Some fifty experimental models were constructed over five years, and in 1645 he had his machine. It was the size of a glove box, simple in appearance, portable and workable. (This very machine is now, as students believe, in private ownership in the south of France.) Pascal hoped to get rich on the invention and staked a claim for a patent; by 1652 a standard

9

model was in production and was placed on sale at 100 livres. His hopes were not realized. The manufacturing costs were too high, and the invention of logarithms cut the demand for the machine. Yet it was the first of all computers; as Gilberte writes in her *Life* of her brother, "He reduced to mechanism a science which is wholly in the human mind." Now Blaise began to be referred to as *le grand M. Pascal.*

His concern with religion was growing, but it was still far from pervading his thoughts. The drama and poetry of the day did not appeal to him. He disliked and distrusted their emotional influence; he found them artificial and stilted. He read in philosophy and was attracted to the ancient Stoic writings and to Epictetus. What he admired was the value they placed on fortitude and duty, the high notion that man's fate is in his own power, that no outside force can crush him. But it was a "fatal flaw," he believed, to make man the complete master, to leave no place for God. Stoicism was too self-assured, too filled with self-pride, too apt "to greet the unseen with a sneer."

When his computer was finished, a new interest came to Pascal. Pierre Petit on a visit to Rouen brought news of Evangelista Torricelli's experiments in Florence two years before, in which he had arrived at the first correct view of the nature of atmospheric pressure. He had demonstrated that we live immersed at the bottom of a sea of air and that air has weight. Taking a glass tube closed at one end, he filled it with mercury, and covered the open end with his finger; when he inverted the tube, placed the open end in a bowl of mercury and removed his finger, the liquid in the tube did not entirely empty into the bowl, but sank only part of the way down the tube and remained mysteriously suspended. "I assert," he wrote, "that the force [holding up the quicksilver] is external and comes from without . . . [and that because of it] the quicksilver enters and rises in a column high enough to make equilibrium with the weight of the external air which forces it up."

This was a remarkable and highly controversial conclusion

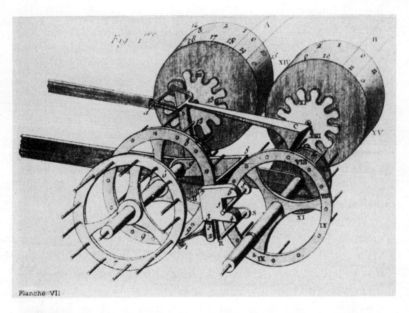

A portion of the interior of Pascal's machine à calculer. *(From a drawing in* L'Encyclopédie Panckouke, *1784)*

because it contradicted what Aristotle, the Scholastics and Descartes had said about the impossibility of a vacuum. When the mercury that had originally filled it sank in the tube, what remained in the vacant space? Nothing, said Torricelli; but, said Descartes, nothing can be filled with Nothing. If any portion of the tube were empty, it would instantaneously collapse. Pascal, excited by Torricelli's discovery, decided to make his own experiments to vindicate and extend it. Rouen had the best glassworks in Europe; Étienne had money to pay for research; Petit had the engineering experience; Blaise had the enthusiasm. With the help of glass tubes of different lengths, breadths and shapes (two were 46 feet long and were bound to

11

ships' masts to strengthen them), he was able to show that the space vacated by the falling mercury was indeed a void. Neither rarefied air nor mercury vapor, as the "plenists" desperately asserted, was to be found in the vacant space. He varied the apparatus; he invented the plunger type of syringe to serve as a vacuum pump; he gave sensational public demonstrations with five-story-high siphons before five hundred of the city's Eminences. All his ingenious experiments gave a consistent result: for a given liquid the height of the column in the tube was always the same; it was unaffected by the space at the top; it was not being sucked up from above but must be pressed up from below and from outside.

He now had in mind the writing of a treatise on the vacuum, and for this purpose other experiments were needed. Momentarily, however, he was diverted by a religious crisis.

It was occasioned by an accident to Pascal's father. On a night in January 1646 Étienne hastened out of doors to prevent a duel, slipped on ice and dislocated his thigh. First aid was given by two "reformed characters," brothers of noble birth whose name was Deschamps, who had been converted by a pious curé of a Rouen suburb and were devoting their lives to good works, "especially the good work of bone-setting at which they were adept." The example of their charity and fervor awakened religious feelings in the whole Pascal family. Each was deeply stirred and each responded in character. Étienne decided the time had come to look to his spiritual progress; Gilberte and her husband became "consistently devout"; Jacqueline moved closer to the decision of giving herself entirely to the religious life. Blaise's response was "more enigmatic." He immersed himself in theological tracts; he began serious study of the Bible. His letters begin to show that he was "aflame with religious zeal." "*Les choses corporelles,*" he wrote Gilberte, "*ne sont qu'une image des spirituelles, et Dieu a representé les choses invisibles dans les visibles.*" Undoubtedly he was in conflict with himself about his scientific interests. They were "morally dangerous" in the eyes of the Deschamps brothers "as leading to intellectual pride and distracting the attention from the quest of salvation," and of dubious value. But the realist and the lover of natural knowledge could not bring himself to renounce his experiments. He had undergone a crisis, had turned in a direction that he would never abandon, and yet the years 1648-54 were to be his most fruitful and triumphant period of scientific discovery.

In 1647 he fell seriously ill, the diagnosis being "overwork." He was medically advised to avoid all intellectual work and to seek relaxation in society. (His physician was obviously a realist.) He retired to Paris with Jacqueline as his housekeeper and secretary. Soon he was back to his experiments and working at his *Nouvelles Expériences Touchant le Vide*. Descartes visited him in Paris and they discussed Torricelli's work. A couple of months later Pascal arranged to have his brother-in-law conduct an experiment on the Puy de Dôme, a mountain near Clermont, to determine whether a column of mercury in a tube would stand higher at the base of the mountain than at the summit. The results were conclusive, and the way was open for hydrostatics and hydrodynamics.

He sketched two treatises, *La Grande Expérience de l'Équilibre des Liqueurs* and *De la Pesanteur et de la Masse de l'Air*, neither of which was published till after his death. The pressure of the air was, he saw, a clue to a more general law of the behavior of "liquids" (including gases). His experiments led him to a startling principle. Pressure, it appeared, exerted at any place on a fluid in a closed vessel, is transmitted undiminished throughout the fluid and acts at right angles to all surfaces. Suppose two tanks of equal size are filled with water and joined by a pipe at the base; evidently the water in the tanks will be in equilibrium like equal weights in the pans of a scale. But suppose one tank is a hundred times as large as the other, what then? The water will still be in equilibrium. In other words, the smaller weight balances the larger. As Pascal

13

explained: by putting into each tank a piston that fits it exactly, "a man pushing on the small piston will exert a force equal to that of one hundred men who are pushing on the piston which is one hundred times as large. . . . It may be added, for greater clearness, that the water is under equal pressure beneath the two pistons; for if one of them has one hundred times more weight than the other, it also touches one hundred times as many parts of the liquid." From this principle, which leads to anomalies and paradoxes that have puzzled students for years, came the hydraulic press.

"Nothing that has to do with faith can be the concern of the reason." This was the principle Blaise had learned in his boyhood. Some might regard it as even more anomalous than the principle of the behavior of fluids, but he lived by it and worked it into greater fullness. He held his religious beliefs with a growing fervor, but he refused to be "stampeded into a flight from reason." In one of the most famous of the *Pensées* he says: "It is the way of God, who does all things gently, to put religion into the mind by reason, and into the heart by grace."

The period from 1648 to 1653 was one of uncertainty and inner struggle in Pascal's life. His father died; his beloved sister Jacqueline, who had been his closest companion, finally renounced the outer world and entered the convent at Port Royal. He contemplated marriage, but not very seriously. Bleak and lonely though his life had suddenly become, he was better off single: like Charles Lamb he was one of "nature's bachelors." He did not, however, cut himself off from the *salons* of Paris society. He was a man of science, but he was also a man of fashion and ambition, besides which he enjoyed "savoring human types." He formed an important friendship with a great nobleman, the young Duc de Roannez, who was, strangely, more interested in science than in women. Pascal lectured to a brilliant assembly in the drawing room of the

14

Duchesse d'Aiguillon, Richelieu's niece, demonstrating his calculating machine and speaking on recent advances in hydrostatics. Mindful, as always, of his business interests, he sent Queen Christina of Sweden — who had recently finished off poor Descartes by requiring him to expound science and philosophy to her on cold mornings before breakfast — a gift of one of his calculating machines. This was accompanied by a celebrated letter, which, though suitably gracious in tone, said in effect: "Your Majesty is a very great person by virtue of your sovereign rank. As we may agree, I am a still greater person by virtue of my sovereign intellect."

His social relaxations did not prevent him from reading and writing. At the very height of this period it is likely he produced, besides scientific essays, a large part of the marvelous *Pensées*, which were not published until after his death. He was also immersed in Montaigne's *Essays*, which filled him with contradictory emotions. He cherished Montaigne's honesty, his hatred of cruelty, his scorn of dogmatism and love of intellectual freedom, his benevolent interest in human nature, "the stoical tranquillity he practised and preached in an evil time;" and he admired the incomparable vitality of his style. But the *Essays* also shocked and horrified him. Montaigne had an ironical and detached view of religion. If religion was sublime, it was also celestial; it was heaven's affair rather than man's. Why should one imagine that municipal laws are the laws of the universe? Man is miserably weak, supremely foolish, moved by base passions and vanity. Wisdom and justice are unknown to him. His sages and philosophers are chatterers. He can rely upon neither his senses nor his sense. There was no vast difference, says Mortimer, between Montaigne's outlook and Hobbes' — that "the life of man is solitary, poor, nasty, brutish and short." Montaigne's unforgettable phrases stirred and seared Pascal. Man, Montaigne said, is the prey of trivial chances. "A gust of contrarie winds, the croking of a flight of Ravens, the false pase of a Horse, the casual flight of

15

an Eagle, a dreame, a sodaine voyce, a false signe, a mornings mist, an evening fogge, are enough to overthrow, sufficient to overwhelme and pull him to the ground." And if it be urged in mitigation that despite man's wretched condition and contemptible character, he has religion, the reply is, yes, he has religion that chance flung in his way: "Man cannot make a worm, but he makes Gods by the dozen."

In Sainte-Beuve's phrase, Montaigne became anchored in the soul of Pascal. He would not follow where Montaigne beckoned, but he could not turn away from him. This dilemma, centered on one man, was the epitome of the great dilemma of Pascal's whole outlook, of his doubting faith, his troubled convictions, his fluctuating commitments.

In 1653 Pascal made a trip to Poitou with the Duc de Roannez. One of the party was the Chevalier de Méré, a member of the great house of Condé, a soldier, scholar, skeptic and "cultivated libertine." To this man the world owes a debt, made up of strangely incongruous parts. Méré was an elegant and skillful writer; Pascal's style, says Mortimer, attained "its complete freedom from affectation and verbiage" from Méré's teaching. The other part of the debt is more intriguing and important. Méré was a gambler, and the occasion arose, shortly after the journey to Poitou, for him to ask Pascal to solve two problems of practical use to gamblers. Pascal had for some months been disaffected and weary, preoccupied with his soul. On his return to Paris, however, he began to settle his estate, to buy property, to renew his efforts to sell his arithmetical machine; and before long his scientific interests, touched off by Méré, "awoke from a long sleep." The first problem was: when playing with two dice, what is the minimum number of throws on which one can advantageously bet that a double six will turn up? This was easy for Pascal; 24 throws would be a bad bet, 25 a good one. The other question was harder.

Two players have agreed that the stakes will go to the one who first wins three games. Before this happens their play is interrupted. How, under different circumstance, are the stakes to be divided? Pascal found an ingenious and simple answer. Suppose A has won two games and B one, and the stake is sixteen pistoles. In that case A should receive twelve pistoles and B four. Pascal reasoned as follows: Assume that one more game could be played. If A won it, he would be entitled to the full sixteen pistoles; if B won it, at that point he could claim half the stakes. So when only three games have been played A can say to B, "Win or lose the next game, I have at least eight pistoles due me. That leaves the other eight for the next game, in which the chances are equal, so if we cannot play the next game we divide equally, and I get four." Pascal worked out the other cases. He sent his solution to Pierre de Fermat at Toulouse, who got the same results by algebraic methods. Pascal was very pleased. "*Je vois bien,*" he wrote, "*que la vérité est la même à Toulouse et à Paris.*" The independent work of both men was a foundation stone of the mathematical theory of probability. (Cardan had anticipated some of their results in the preceding century, but his little book on the subject lay unnoticed.)

Pascal's success in solving Méré's problems stimulated Pascal to further study of the mathematical theory of chances. One of his brilliant ideas was the arithmetical triangle, a capsule of the calculus of probabilities. He flung himself into this work with immense energy; it was his way when his interest had been roused to tackle a subject — physics, theology, mathematics, even business — in a frenzy, as if nothing else mattered, as if time was running out. His program was unbelievably ambitious. He proposed to write treatises on the theory of numbers, on the equilibrium of liquids, on the arithmetical triangle; to write papers on magic squares, on circles, on conic sections, on perspective; and "to reduce to an exact

17

art, with the rigour of mathematical demonstration, the incertitude of chance, thus creating a new science which could justly claim the stupefying title: the geometry of hazard."

Of this fantastic plan only a small part came to fruition. Its grandiosity marked his desperation. He had to keep going, to drive himself, whether in high society or high science, so as not to be alone with his dark thoughts. He was "a made man and a celebrity"; yet he was a tormented man. He doubted his God; he felt empty and lost.

On November 23, 1654, he experienced another intense religious episode, "a timeless eternal moment." Immediately after his vision he wrote at headlong speed an account of it, which, together with a parchment copy, was found after his death, sewn into his doublet. His life from then on was changed. Prayer and sacred study were his main concern. He never again attached his name to any of his writings ("an evident discipline against vanity"); he withdrew more and more from society. His health began to deteriorate. For a brief period he entered a retreat of the Port Royalists. Doctrinal questions fevered him. We are "tied and bound with the chain of our sins," but does this mean there is no hope, that the truth is beyond us? He thought not; there is a function for the human will, a way to salvation; it is not as prideful as the way of Epictetus, nor as limited as the way of Montaigne. Out of the quarrel between the assignment of moral power and moral weakness he drew a synthesis: "Faith teaches us to assign these two inclinations to different things: infirmity to human nature, but power to grace."

I shall not follow him in his career as a pamphleteer. His famous anti-Jesuit *Provincial Letters* are a mixture of tedious theological claptrap, moral grandeur, humor and irony. They are masterpieces of polemical literature: even Voltaire, who despised everything about Pascal, found the wit of the letters a match for Molière's finest comedies. The *Provincial Letters*, the background of the dispute that occasioned them, the history of Port Royal, and Pascal's relation to Jansenism are all well described in Mortimer's excellent book.

In the years that remained, though religious questions were uppermost in his thoughts, Pascal neither entirely abandoned science nor renounced fashionable society. He wrote his *Esprit de Géométrie*, a philosophical essay, which, in Mortimer's view, is Pascal's equivalent of Descartes' *Discourse on Method*. He greatly advanced knowledge of the mathematics of the cycloid. One night, awakened by a violent toothache, there came, as his sister Gilberte tells it, "uninvited into his mind some thought on the problem of the roulette [i.e., cycloid]." A whole crowd of thoughts followed: he determined the area of a section produced by any line parallel to the base, the volume generated by it revolving around its base, and the centers of gravity of these volumes. Anonymously he sent out a challenge to all mathematicians to solve these problems; entries came from Christian Huygens, Christopher Wren, John Wallis and others. No one matched Pascal's solutions, which, when published, caused a sensation, not only as an intellectual feat but also because he stirred old and bitter scientific quarrels. He did not attach his name to the publications, but there was no doubt who was the author.

Now at last — this was 1659 — he "bade a final farewell to the glories and quarrels of science." This may be the moment "at which he began to wear next his skin an iron belt with small spikes which he pressed at any temptation to pride." As his health worsened, he became a solitary and an ascetic. He gave up his carriage and horses, his tapestries, even his books, except for the Bible "and (surely) Montaigne." He found it difficult to eat and could take only liquids.

He wished once more to meet with his friend Fermat, but he was not strong enough to make the journey even to a midway point. His letter to Fermat, explaining why he could not come, gives a moving self-portrait: "I find geometry the noblest exercise of the mind, yet I know it to be so useless that I

19

see no difference between a geometer and a clever artisan. I call it the loveliest occupation in the world, but only an occupation. . . . A singular chance about a year or two ago did set me at mathematics, but having settled that matter I am not likely ever to touch the subject again."

He was dying. He had resigned himself and was waiting upon God, filling a page of the *Pensées* when he could. Yet in the midst of his departure came an almost comic interlude. Suddenly he immersed himself in the formation of the first omnibus company, which soon ran its first line from the Porte Saint-Antoine to the Luxembourg. It was a far more profitable venture than his calculating machine. He made money, which he gave to charity, and bequeathed half his interest in the company to the hospitals of Clermont and Paris.

In June 1663 he was attacked by terrible headaches. Convulsions followed. On August 18 he died, aged thirty-nine.

Mortimer has made Pascal understandable. But perhaps, after all, "realist" is not the best word for him. He had many gifts; he was many things. He was divided as experience itself is divided; he was uncertain in an uncertain world. His scientific achievements were extraordinary, yet it is in the *Pensées* that he achieved supremely. He had a profound sense of man's loneliness, of his terror of nothingness and longing to find his place in a vast, indifferent universe; and an exquisite way of transfixing truth, of making anguish stand still, of speaking the questions that men have asked since the beginning. No one, neither scientist nor philosopher, neither rationalist nor mystic, has transcended his insight into man's condition:

"For after all what is man in nature? A nothing in relation to infinity, all in relation to nothing, a central point between nothing and all, and infinitely far from understanding either. The end of things and their beginnings are impregnably concealed from him in an impenetrable secret. He is equally incapable of seeing the nothingness out of which he was drawn and the infinite in which he is engulfed."

20

JOHN STUART MILL

J OHN STUART MILL, economist, philosopher, social reformer and child prodigy, was one of
the remarkable men of the nineteenth century. How remarkable and how interesting was perhaps not fully realized until
the publication of an excellent biography by Michael St. John
Packe, a history scholar of Magdalene College, Cambridge.*
It may seem surprising that there are fresh things to be said
about Mill. He has lacked neither biographers — among them,
the noted Scottish philosopher and psychologist, Alexander
Bain — nor admirers; he was enormously influential in economic, political, social and philosophical thought, so that even
today, 88 years after his death, few names are more familiar

* Michael St. John Packe, *The Life of John Stuart Mill*, New York, 1954.

than his to educated persons; he wrote a celebrated autobiography that tells a great deal about his mental development and intellectual life. Yet to a large extent the man himself remained veiled and forbidding. There was much about him that was obscure — about his personality, his friendships, his famous, long-drawn-out love affair with Harriet Taylor, which for twenty years or more nourished the gossip of London society. The source materials that could shed new light on these matters and fill in many gaps of information were long withheld from public scrutiny by Mill's stepdaughter and her niece, and were later scattered when offered at auction. Within the last twenty years, fortunately, these materials were reassembled, and with their help Packe has undertaken a full-scale portrait. I cannot say that he has made plain everything that was not plain before, nor are all his conclusions and appraisals equally compelling. But he has written a fascinating, sensitive book, which brings into full view a hitherto half-hidden figure. Packe's Mill is a man who moves you, whom you like and dislike and pity as well as admire. It is too much to ask that he, of all men, be made wholly understandable.

The prodigiousness of prodigies blinds us to their other traits. What lives inside the marvel is apt to be forgotten. But Mill the man is inexplicable if one fails fully to take into account the incredible child and its even more incredible childhood. His father, James Mill, was the eldest son of a Scotch shoemaker and of an energetic woman who had determined that her first-born should be well educated and rise in the world. James did not disappoint her. A man of ability as well as self-confidence and resolution, he fought his way up. After making the most of as good an education as Scotland could provide, he moved to London, became a journalist, married a pretty girl with a dowry of £400 and undertook to write a *History of India*, a task for which he considered himself especially well qualified because he had never been to India and

John Stuart Mill in the last year of his life. Painting by G. F. Watts.
(The Bettmann Archive)

23

therefore "would be the better able to take an objective view of the whole vast field." In 1806 he produced a son — John Stuart — and in 1808 he met Jeremy Bentham. The second event profoundly influenced the consequences of the first. Bentham, whose great work, *Principles of Morals and Legislation*, had appeared in 1789, was the founder of utilitarianism, a theory that held as the *summum bonum* the greatest happiness of the greatest number. Happiness, he said, was to be measured by pleasure — though it must not be supposed everyone would have enjoyed what he enjoyed. He believed also that men could be made more virtuous by virtuous legislation and that there was no limit to the benefits a good education could confer. James Mill became Bentham's disciple and close friend, and in John Stuart, though he was only three, both men hoped to find not only another worthy disciple of their cause but an ideal experimental organism for testing their theories as well.

"*L'éducation peut tout,*" said Helvetius: it was upon this premise that James set out to educate his son. The tender mind, James held, was a blank sheet upon which the teacher could write what he liked; the child's experiences arranged themselves in patterns from which inclination and habit were formed — like Bentham, Mill subscribed to associationist theories of psychology — and so by regulating the flow of experience, character and ability could be made to order. At three the boy began to learn Greek. The words were written out on cards together with their English meanings; when he had mastered this vocabulary, John was introduced to Aesop's fables. Next came Herodotus, then Xenophon, and at age seven he read Plato's *Dialogues*, which he later felt "would have been better omitted, as it was totally impossible I should understand [them]." James was not the most patient of men; moreover, "he demanded of me," wrote John, "not only the utmost that I could do, but much that I could by no possibility have done." Yet if the son was not spared, neither was the

24

father. There was nothing to which he would not submit to further the child's education. He told the boy what to read, heard his lessons, discoursed with him on long walks, drilled him in arithmetic. In the evenings they sat together on opposite sides of the table in the crowded living room, with one baby sister "howling in one corner" and another in another, John studying his Greek, James doing his journalistic chores or grinding away at his *History of India*. As there was in those days no Greek-English lexicon, John incessantly interrupted his father to ask the meaning of words. It is not surprising that the *History*, which was to have taken three years, took ten. What is remarkable is that it got done at all and that James did not bash in the head of the diminutive scholar.

But here, for once, theory came to John's aid. For neither Bentham nor James Mill believed in the use of force in education. Children were to be persuaded, reasoned with, trained to perceive their true interest by the skillful administration of pleasures and pains so that in future life the prudent course would be followed almost by instinct. Ranting and displays of choler by the teacher were permitted; also sarcasm; but the laying on of hands or more abrasive instruments was strictly forbidden, an extraordinary departure "at a time when the whack-happy Dr. Keate was the *dernier cri* at Eton."

The system soon began to pay off — at least in respect of what was intended. John started reading on his own: biographies, accounts of voyages and explorations, the histories of Hume, Gibbon and Robertson, political works, the plays of Terence, the *Iliad* and the *Odyssey*, a translation of Plutarch, Aristotle's *Rhetoric*. At five "he was able to engage Lady Spencer, wife of the First Lord of the Admiralty, in an animated comparison of Wellington and Marlborough"; at six and a half he composed a Roman history in fifteen hundred words. While the danger of his becoming a dull boy seemed small, books of amusement were not denied him. *Robinson Crusoe* was his chief delight; and his father went so far — his

25

own library being deficient in such giddy items — as to borrow for his son the *Arabian Nights*, *Don Quixote*, Maria Edgeworth's *Popular Tales* and a "book of some reputation in its day," Brooke's *Fool of Quality*.

In John's eighth year "the pulse of his learning quickened." He added Latin, algebra and geometry to his curriculum. He raced through Cicero, Horace, Livy, Virgil, Ovid and Juvenal. Before he was twelve he had pierced the mysteries of trigonometry, conic sections, the differential calculus and other portions of higher mathematics. In this field he was mostly on his own, because his father was not much of a mathematician. This, of course, did not deter him from urging the child on or scolding him repeatedly for stumbling over difficult problems for the solution of which he lacked the necessary knowledge. John continued avidly to read history — Mitford's *History of Greece* was a favorite — and to compose several more ancient histories, including a full-scale analysis of Roman constitutional law, all of which he later destroyed "in contempt of my childish efforts." His father introduced him to the pleasures of poetry with Pope's translation of the *Iliad*. He read it again and again, attempted to write verse in the same style, turned to Greek poetry, then to Milton, Goldsmith, Gray, Cowper, Spenser, Scott, Dryden, Burns. James was convinced that the best way of developing and practicing knowledge was to teach others what had been freshly learned. Accordingly John was expected to coach his little sisters and brothers, of whom in time there were eight. When he was eight and just beginning Latin, his sister Wilhelmina, aged five, became his pupil in classics, and he was responsible for the answers to questions on Latin grammar put to her by her father. Wilhelmina was not retarded; she read Cornelius Nepos when she was six, and dabbled in Caesar; the other children kept pace — from Ovid to cube root — but Packe warns against concluding they were erudite. In fact, "in later life they were under the impression that their own education had been somewhat sketchy."

While the system did not rule out learning by demonstrations, James tended to be "over-satisfied with abstractions." Chemistry became one of young John Mill's greatest amusements, but it was the chemistry to be found in books, and it was years before he actually witnessed an experiment. In any case his father preferred that John spend his time on economics, logic and political theory. When he was thirteen he was taken through "a complete course of political economy," including lectures on Ricardo, delivered by James on their walks in the country. The intricate subject of money was carefully gone into, and lessons in logic became part of the daily fare.

In 1820, when John was nearly fourteen, his lessons came to an end. His father had taught him all he could, and it was now decided that "the only thing left for him to learn was how to take his place with other human beings." Even James could see that this would not be easy. His system had required that John be quarantined from other boys lest they corrupt him with their "vulgar modes of thought and feeling." (So he says in the *Autobiography*.) He was healthy and hardy, but he played no games and was too awkward to learn. He had been trained, as he tells us, to *know* rather than to *do*. He was not shy; he was inclined to be "sharply dogmatic in his statements"; yet he did not regard himself as more gifted than others. In his later self-appraisal he rated himself as "rather below than above par" in quickness of apprehension, retentiveness of memory, energy and initiative: "and if I have accomplished anything, I owe it, among other fortunate circumstances, to the fact that through the early training bestowed on me by my father, I started, I may fairly say, with an advantage of a quarter of a century over my contemporaries."

It was arranged to expose him to the outside world by sending him to France to live for six months with Bentham's brother Samuel, who had a family of children somewhat older than John. The trial was successful beyond expectation. At

27

first he continued the routine of improving his mind, learning French, reading Racine and Voltaire, tackling Legendre's *Éléments de Géométrie*, writing long letters to his father about current happenings such as the French elections. But the Benthams ran a relaxed, cheerful, agreeably disorganized household, and in time John adapted himself to its ways. He went on picnics, attended peasant dances, visited the circus, looked through a telescope, became interested in botanizing. A French tailor fitted him for more fashionable clothes; a gentlewoman taught him piano. Fencing, riding and dancing were added to his accomplishments, though it cannot be said he became accomplished in these pursuits, much less cared for them. By and by he cut down on his mathematics and caught more butterflies, studied less logic and practiced more sleeping. Once, it is said, he stayed abed till nine o'clock. The "free and genial atmosphere of continental life" — so characterized in the *Autobiography* — warmed him. Six months stretched into a year. When he returned home in July, 1821, he was not the same boy. To be sure, he had not lost his interest in intellectual pursuits, and he continued his reading in logic, metaphysics, psychology, and political economy. But he had glimpsed new sights that thrilled him, learned to live outside the tiny, confining circle of his childhood, seen in their daily habit the working folk hitherto known to him only "within the graphs of political economy." The wonder child had outgrown its test tube.

John's godfather, Sir John Stuart, had left £500 to send him to Cambridge, but James refused to allow it on the ground there was nothing Cambridge could teach his son. He was probably right. When John was sixteen he accepted an invitation to speak before the Cambridge Union. "His massive power in disputation, uttered from a flimsy body in the creaking tones of sixteen, stilled the brittle oratory of the adolescent giants. He left a great impression." The same winter he founded a society of his own. He gave it the name Utilitarian

Society, and its five members met fortnightly for three years, to debate serious matters. In 1823 he was allotted a post as a junior clerk with the East India Company, writing dispatches to the three Indian Presidencies on political subjects. Here he was to remain for thirty-one years until his retirement in 1858, rising slowly to Chief Examiner, the highest post in his department, and very well paid. Thus financial security, like almost everything else, came to him early.

Another event, which might be said to have completed his education, occurred in his seventeenth year. He was arrested for distributing "obscene literature." The circumstances did him credit. The famous reformer Francis Place, while accepting Malthus's arithmetic, was not disposed complacently to accept his "natural" correctives to overpopulation: misery, vice and war. And as a practical man he had misgivings as to the effectiveness of moral restraint. Having been told that "the French manage these things better," he proceeded to Paris and learned what was to be learned. Forthwith he incorporated this knowledge in a pamphlet titled "To Married Working People" and had it distributed in a broadsheet. One summer morning, walking to his new job at India House, young Mill found in St. James Park a baby, "blue, new-born and strangled, wrapped up in grimy rags, and left." This did not seem to him, any more than to Place, the solution to the population problem. His sense of horror was that same morning reinforced by the spectacle of criminals dangling from the gibbet at the Old Bailey. He told Place what he had seen, was given some broadsheets and, with a friend, undertook to strew them through London. The boys were soon caught, locked up and brought before a magistrate. He remanded them to the Lord Mayor. This august official lectured them on their outrageous attempt "to corrupt the purity of English womanhood," sentenced them to fourteen days and then let them go after a day or two.

Years later, when he had grown famous, Mill met the Lord Mayor at a banquet. "The Lord Mayor beamed civilly; 'I have had the pleasure of sitting opposite you before, Mr. Mill,' he said. Mill agreed tartly: he would have a happier memory of the occasion, he replied, if the Lord Mayor had been as quick then as he was now in perceiving opposites: for he would have been able to discriminate between an attempt to prevent infanticide and the promotion of obscenity."

It is an outstanding merit of Packe's book that it makes clear how much Mill owed to his father, to the moral and intellectual environment of his youth. Thus is corrected a misimpression for which Mill himself was largely responsible. The *Autobiography* depicts his childhood as "a weary drudgery"; his home as "cheerless, godless, silent and afraid." But Packe shows us much that alleviates this unbrokenly gloomy picture, and it becomes even harder to accept on contemplating not only Mill's works but also his character traits, his ethical outlook, his affirmations of social idealism.

Mill's early training is reflected in his self-discipline and self-directedness. He never failed to finish what he started, he had an enormous capacity for work and he was usually engaged in a dozen undertakings besides his regular job at India House. He mixed in politics, stood for Parliament and was elected. Articles, reviews, polemics, essays flowed from his pen on subjects ranging from the libel laws to the theories of Coleridge, from the economics of labor to the metaphysics of Sir William Hamilton, from contemporary affairs to Tennyson's poetry. The journals to which he contributed sometimes paid him for his work, but often it was he who kept the journal going by digging into his pocket. Mill was unfailingly generous; he never forgot the examples set by Bentham, Place and others who had saved James Mill and his family from poverty. Though he was far from wealthy, he gave money freely to charitable organizations, to workingmen's libraries, to the Woman's Suffrage Society, to "a Wiltshire poacher under prosecution," to girls' schools, to the Drinking Fountain Association, to indigent writers and philosophers of every shade of opinion. That prickly, hapless, half-madman, half-genius, Auguste Comte, gained from Mill's generosity, as did also the even more offensive Carlyle.

It may be urged, of course, that whatever Mill did for Carlyle was paltry in light of the latter's magnificent equanimity in the famous affair of the burned manuscript. He had lent Mill the completed manuscript of the first volume of his *French Revolution*, the product of five months' arduous literary labor. Mill took it home, read it, and then with incredible addlepatedness swept the sheets, together with some old papers he was clearing out, into a bin for "kitchen use." The stove consumed the lot, leaving only a few charred pages. Carlyle, who had destroyed even the notes of his work before sitting down to write, as was his habit, carried off the catastrophe with unexampled generosity and understanding. "I feel," he wrote Mill, "that your sorrow must be sharper than mine: yours bound to be a *passive* one."

But it was more than his generosity, honesty and scrupulous fair dealing that brought Mill as an adult the friends of which he had been deprived as a child. At the age of twenty he fell into a profound mental depression, suddenly discovering that he was incapable of feeling. After some months the crisis passed: "I was no longer hopeless," he wrote, "I was not a stick or a stone." The experience brought a change in outlook. Thenceforward he gave to his sympathies and passions a greater role. He recognized that he needed companionship, and he sought it. He mixed more in society. George Grote, the historian, Alexander Bain, the psychologist, John Sterling, the poet and author, were among those who became his close friends. In time he also fell in love. His father, to be sure, had loved him, but in his fashion: James Mill was an icy man; in

31

his normal mood he was astringent, otherwise irritable. As for John's mother, she was a crushed creature who bore and raised children, though she seems to have had only a small share in the raising of her eldest son. John felt sorry for her; later his compassion turned to scorn. There is not a single reference to her in the *Autobiography*. Yet it must not be inferred that he looked down on women generally. He was not only "fanatically devoted" to their cause but capable of a transcending passion for one woman.

In 1830 he met Harriet Taylor. She was young, slender, tiny, beautiful, poetic, philosophical, passionate, proud and clever. Carlyle described her as a "living romance heroine, of the clearest insight, of the royalest volition, very interesting, of questionable destiny. . . ." When he had grown to hate her, he drew a no less vivid sketch, observing that "she affects, with a kind of sultana noble-mindedness, a certain girlish petulance." This was a minor matter. More distressing was the fact that when Mill met her she was already married. John Taylor was a prosperous London merchant, who adored his wife. He was destined to become the most long-suffering husband of history, denied even the satisfaction of being forthrightly cuckolded. The instant Mill laid eyes on Harriet "a passion sprang out of the bushes like a hundred Ashantees, and he was carried away captive." For twenty years they were lovers, and it was not until 1851, two years after John Taylor won release from his ridiculous situation by death, that they were married; but if all the biographers, including Packe, are right, they were lovers living for two decades in a state of "technical innocence." Harriet was the kind of girl who knew how to manage her relations with two men, giving neither exactly what he wanted but each enough to make him reluctant to give up what he had. Taylor first tried to break up the affair. When Harriet outtalked him, he resigned himself to the wretched, lonely life of a husband of convenience. Harriet spent a good part of each year in her country house or traveling in Europe with Mill, and even when she lived with her husband in London, Mill came twice weekly to dine and Taylor trotted off to his club to spare the lovers' feelings.

Mill knew that divorce was impossible and settled for inspiration and companionship. Packe tries hard to convince us that Mill was not undersexed, and that while he had no respect for the marriage institution, he was quite prepared to deny the appetite of the flesh for the higher spiritual pleasures of abnegation. "Mill had the Victorian's passion for liberating the human mind from everything including sex," an English reviewer had remarked. I do not find Packe very enlightening or original on this aspect of the Mill-Taylor affair. There comes to mind Carlyle's brilliantly disagreeable question on the occasion of one of John's trips abroad, accompanied by Harriet, seeking to recover his health: "They are innocent says Charity, they are guilty says Scandal: then why in the name of wonder are they dying brokenhearted?"

Packe does, however, succeed in making abundantly clear how profoundly Harriet influenced Mill's views. Within a few years after they met, their companionship had become as close intellectually as emotionally. He extended her horizons and she passed meticulous judgment on his ideas. "From 1846 onwards," Packe writes, "Mill's entries in his chronological list of published works are marked with increasing frequency as a 'joint production with my wife,' and the *Political Economy* is so styled. The influence she had gradually extended over him now ended in complete ascendancy, and his further writings were 'not the work of one mind, but of the fusion of two.' " It was a good influence; of this there can be no question. John Mill was a reasoning man, a scientific and "severe" thinker, without either much sense of the concrete, as his friend John Sterling said, or much poetry. Harriet supplied these qualities to his work; it was she who never permitted him

33

to relax his vigilance against social injustice, who brought him to realize, he said, the practical possibilities of socialism, who "helped to keep him a radical and a democrat."

Mill's best-known writings include the *Logic*, the *Political Economy*, and the famous essay *On Liberty*. Packe scarcely gives more than the gist of Mill's writings and offers no critique in any real sense. *A System of Logic* has as its main object the "rehabilitation of induction." Mill was a mechanist. He shared the nineteenth-century confidence in the final reducibility of the most diverse phenomena to mechanical actions. He adopted as a grand hypothesis the uniformity of nature, and as another major assumption the "Law of Universal Causation," which holds that "every event, or the beginning of every phenomenon, must have some cause, some antecedent, on the existence of which it is invariably and unconditionally consequent." On this base rests the inductive method whose regulating principles or "canons" "provide a test of the validity of inductive reasoning, similar to the syllogistic test of ratiocination." Since the causes of an event are seldom plain and simple, and since experiment and observation are beset by confusions, chimeras, hidden factors, Mill hoped that the canons would guide scientists along dependable paths to safe summits. It is interesting to observe that he went so far in rejecting the notion of absolute or necessary truths espoused by the intuitionist philosophers that he dismissed the conclusions of geometry as mere deductions from arbitrary assumptions that are neither necessary nor true (e.g., "There exist no points without magnitude; no lines without breadth, nor perfectly straight . . ."). As for the rest of mathematics, he held it to be founded on the science of numbers, which consists only of "experimental laws based on the evidence of the senses."

The *Logic*, like the *Political Economy*, was an immensely successful and influential book. At Oxford and Cambridge it

34

was for a half century the "groundwork of natural science." Its shortcomings, which are serious, arise in part from the fact that Mill "claimed for reason more than it could achieve," in part from the inadequacies of his notions of psychology. He hoped, of course, to apply his principles of logic to man and society, always of greater interest to him than the sciences themselves, and the concluding portions of his book are devoted to these considerations. But here he was soon mired in classical dilemmas and difficulties. It was all very well to say that human behavior was a phenomenon subject to causal laws, like the behavior of a gas or the revolution of a planet; but where did this leave the freedom of the human will? Mill was unable to resolve the dilemma without, as Packe says, bending his logic. The results were not satisfactory, but the attempt is noteworthy in demonstrating once more that Mill, though a worshiper of reason, was "first of all a humanist who carried the stern standard of utilitarianism into philosophy itself." He would not accept, whatever the cost to intellectual consistency, conclusions "repugnant to the interests of mankind."

And so again in the essay *On Liberty*, it is "more of a hymn or incantation" than an impeccably reasoned discourse. Written jointly with his wife, informed with her ideas, cloaked in his eloquence, the essay perhaps more than any other of Mill's works retains its stature today. Its outlook is eclectic. Mill was incapable of glossing over the dilemmas of freedom. The individual's rights and the needs of society, majority rule and minority opinions, "the tyranny of a few" and the "rough justice of the many," the evils of collectivism and the immorality of *laissez-faire* — these were the conflicts and incompatibles he could not reconcile and would not deny. On one point, however, he was clear. Both morality and self-interest required the encouragement of minority opinions, the preservation of individual differences. It was important, he advised Bertrand Russell's mother, Lady Amberley, to "establish a character for strangeness." "There is nothing like it," he said,

35

"then one can do what one likes." But individuality was not likely to thrive under the rule of the greatest number. He had read and admired de Tocqueville's *Democracy in America* and he had taken to heart its warnings; and even as he worked for the advancement of the mass of men, for the enlargement of their power, he feared the accompanying pressures toward conformity. "The worth of a State, in the long run, is the worth of the individuals composing it; and a State which . . . dwarfs its men, in order that they may be more docile instruments in its hands even for beneficial purposes will find that with small men no great thing can really be accomplished; and that the perfection to which it has sacrificed everything, will in the end avail it nothing, for want of the vital power which, in order that the machine might work more smoothly, it has preferred to banish."

The "finishing governess" is what Disraeli contemptuously called Mill when he first saw him speaking in Parliament. Disraeli was right, and other unflattering things could justly have been said of Mill. He was cold and egotistical; he was capable of great cruelty to his mother and sisters, whom he despised; he was not a philosopher of the first rank; he was a codifier rather than an innovator. Nevertheless he was a large, admirable figure — courageous, independent, unswerving in his dedication to human welfare and the principles of liberalism, always stately and elevated in his outlook on public questions, one who used his superb intellect only for good. We have need of such finishing governesses.

BERTRAND RUSSELL

I T MUST BE accounted remarkable that no full-scale life has yet been published of Earl Russell.* Too many biographies are written in our day about too many men; even German generals and small-beer statesmen are the subjects of literary portraits. But by the most discriminating standards this distinguished man deserves a biographer. Russell is the foremost living philosopher and a profound logician who with Alfred North Whitehead created *Principia Mathematica*. He is an uncommonly lucid writer. In the absence of a Nobel prize for philosophy or mathematics, he re-

---

* Since this essay was written there has appeared a brief biography of Russell by the late Alan Wood (*Bertrand Russell: The Passionate Skeptic*, New York, 1958).

ceived the prize for literature. Two generations have been instructed by his ideas, delighted by his wit, stirred by his independence, and, not infrequently, spanked by him for their lack of it. Not all creative thinkers lead interesting lives. It may be there is not much more to say about Immanuel Kant than that he was punctual, and about Willard Gibbs than that he rode regularly with his sisters in a carriage around the block. Russell's case is different. He has had an eventful life. He has cared about many things, education and physics, political power and the philosophy of marriage, ethics and relativity, Bolshevism and the foundations of mathematics. He has not only thought his thoughts but also lived them.

Russell has told us about himself. A few years ago he wrote a sparkling essay called "My Mental Development." It was only twenty pages long, but it imparted a good deal of information about the growth of his ideas. In 1957 he published a group of his essays, several of which are autobiographical.* He followed this two years later with an account of his philosophical development.† These works confirm the impression that his life has been as fascinating as it has been fruitful. He has written, as I understand, a full autobiography, but by his direction it is not to be published until after his death. For the present he edifies us with glimpses and fragments.

He was born on May 18, 1872, the younger son of Viscount Amberley, and the grandson of Lord John Russell, the liberal statesman who introduced the famous Reform Bill of 1832. Russell's parents died before he was four years old, and he was brought up by his paternal grandmother at Pembroke Lodge in Richmond Park, the home Queen Victoria gave to Lord Russell. Countess Russell was a Puritan and the habit of her home was austere. Young Russell had to practice piano

---

* Bertrand Russell, *Portraits from Memory and Other Essays*, New York, 1956.
† Bertrand Russell, *My Philosophical Development*, New York, 1959.

Bertrand Russell at the age of 77.
(Photograph by Karsh, Ottawa, copyright)

every morning between 7:30 and 8 before the fires were lit. Then came prayers. "Cold baths all year round were insisted upon." Food was simple, but "if it was at all nice" — apple tart, for example — it was considered "too good for children" and Russell would be served rice pudding. Except possibly for indulging in apple tart, the Countess was no easier on herself. She would not sit in an armchair until the evening, viewed alcohol and tobacco with disfavor and prized only virtue. She must have been an admirable woman. She was unworldly. "She had that indifference to money which is only possible to those who have always had enough of it." She wished for her children that they live useful and decent lives, not that they achieve "success" or marry "well." Above all she believed in private judgment and the "supremacy of the individual conscience." These values were deeply implanted in Russell.

Besides this spiritual legacy, he inherited from his grandparents the genes of longevity and good health. Countess Russell lived to be over eighty; a great-grandmother of Russell's, who was a friend of Gibbon, lived to the age of ninety-two and "to her last day remained a terror to all her descendants"; his maternal grandmother had seventy-two grandchildren, was a founder of Girton College and, after eighty, when she found difficulty in getting to sleep, used to read popular science from midnight to 3 A.M. The only one of Russell's remoter ancestors who did not live to a great age "died of a disease which is now rare, namely, having his head cut off."

The house in Richmond Park was lonely for a child. There were no other children to play with, and Russell's education until he was eighteen came entirely from governesses and tutors. While he did not miss what he did not know, he became, as he describes himself, "a shy, priggish, solitary youth." There was rebellion in him, directed mainly against the theological opinions of his family. They forbade him to read the books in his grandfather's library, so he read them and became interested in history. At the age of eleven he discovered Euclid, "a great event in my life." Mathematics was suspect "because it has no ethical content." He grew to love philosophy, which his family "profoundly disapproved." His intellectual tastes were made, one might say, by opposites. He had a disconcerting way of looking for proofs of things that grownups simply asserted. He recalls that he was told when he was five that the earth is round. He refused to believe it. Thereupon the vicar of the parish, who was Whitehead's father, was called in to persuade the boy. After listening to clerical authority he decided to experiment by digging a hole "in the hopes of emerging at the antipodes." When they told him this was useless, his doubts revived.

His philosophical development, he tells us, began at the age of fifteen. It was dominated by mathematics, but, he writes, "the emotional drive which caused my thinking was mainly doubt as to the fundamental dogmas of religion." The specimens he offers of his philosophical reflections, recorded in a notebook at the age of sixteen (he used Greek letters and phonetic spelling "for purposes of concealment"), are extraordinarily sensitive and moving. His growing disbelief — first in free will, then in immortality and finally in God — frightened and confused him, for these were the teachings of his youth. "I do not think," one entry reads, "it [the search for truth] has in any way made me happier; of course it has given me a deeper character, a contempt for trifles or mockery, but at the same time it has taken away cheerfulness and made it much harder to become bosom friends and, worst of all, it has debarred me from free intercourse with my people, and thus made them strangers to some of my deepest thoughts which, if by chance I do let them out, immediately become the subject for mockery which is inexpressibly bitter to me though not unkindly meant." But neither then nor afterward would such painful experiences deter his desire to know and "to clear away muddles."

At the age of eighteen Russell went to Cambridge. A new world opened for him. He could speak his mind and ask irreverent questions. He was not "stared at as if he were a lunatic" nor "denounced as if he were a criminal." The university had a number of eccentric dons, another group who were competent but dull, and a small class of exceptionally gifted thinkers and teachers. One of the Fellows had the amiable habit of putting a poker in the fire and, when it became red-hot, running after his guests with a view to murder. No one seemed to mind this peculiarity because, owing to a game leg, he never caught the persons he was after. Besides, he was a charming man and was only roused to fury when someone sneezed. Russell's mathematical coach went mad but none of his pupils noticed it. At last he had to be shut up.

Russell's favorite dons included Sir James Frazer, author of *The Golden Bough*; Sir George Darwin, the mathematical physicist; Sir Robert Ball, another noted mathematician; Sir Richard Jebb, the great Greek scholar; and the philosophers Henry Sidgwick and James Ward. Sidgwick was known for having one joke in every lecture. The students waited for it, and after it was told "they were inclined to let their attention wander." He had a stammer, which he used effectively. A German learned man once said to him, "You English have no word for *Gelehrte*." "Yes, we have," Sidgwick replied, "we call them p-p-p-p-prigs."

Whitehead, who was already a fellow and lecturer, examined Russell for entrance scholarships. He took an immediate interest in the younger man and told the "cleverest undergraduates" to look out for him. The result was that in his very first term he met many talented Cantabrigians, some of whom became his lifelong friends. ("I never again," he writes, "had to endure the almost unbearable loneliness of my adolescent years.") Among his closest associates over the years were the philosopher James McTaggart, a Hegelian who shaped Russell in his earlier ideas; Lowes Dickinson, who made his reputation as an author; Roger Fry, noted afterward for his art criticism; the novelist E. M. Forster; the essayist Lytton Strachey, and John Maynard Keynes. In Russell's third year he met G. E. Moore, who was then a freshman. Moore moved everyone who ever knew him by his qualities of mind and character. He fulfilled Russell's "ideal of genius." "He was in those days," Russell writes, "beautiful and slim, with a look almost of inspiration, and with an intellect as deeply passionate as Spinoza's. He had a kind of exquisite purity. I have never but once succeeded in making him tell a lie, and that was by subterfuge. 'Moore,' I said, 'do you always speak the truth?' 'No,' he replied. I believe this to be the only lie he has ever told."

Russell spent the first three years at Cambridge studying mathematics and the fourth studying philosophy. He became deeply interested in the foundations of mathematics, the subject to which he was to make his greatest contribution. He owed much of his inspiration to Whitehead, who showed him many kindnesses, guided him in his transition from a student to an independent writer, and later became his collaborator "on a big book no part of which is wholly due to either."

When his student years ended, the happiest probably in his life, Russell was uncertain whether to follow philosophy or politics. For a time he had a job in the British Embassy in Paris, but then, despite the heavy pressure his family brought to bear on him, the "lure of philosophy proved irresistible." He married, spent some years traveling and visited the United States in 1896. In 1898 he returned to Trinity College as a fellow. Among his earliest writings were *German Social Democracy* (1896), *An Essay on the Foundations of Geometry* (1897), and *A Critical Exposition of the Philosophy of Leibniz* (1900). The *Essay* he now dismisses as containing nothing

43

valid; of a mathematical paper he remarks that it was "un-
mitigated rubbish"; of other early efforts, that they were
"complete nonsense." His mother-in-law, a famous and force-
ful religious leader, assured him that philosophy "is only difficult because of the long words it uses. I confronted her with
the following sentence from notes I had made that day: 'What
is means is and therefore differs from *is*, for "*is* is" would be
nonsense.' It cannot be said that it is long words that make this
sentence difficult." But as time went on he ceased to be troubled
by such problems.

Gradually he weaned himself from Hegel's metaphysic and
embraced empiricism. "The world," he says, "which had been
thin and logical, suddenly became rich and varied and solid."
Russell marks the year 1900 as the "most important year in
my intellectual life." With Whitehead he went to the International Congress of Philosophy in Paris and there heard the
great Italian mathematician Giuseppe Peano lecture on his inventions in symbolic logic. The precision of his discussions
and the power of his elegant notation led Russell to believe
that problems in the foundations of mathematics hitherto ob-
scured in philosophical vagueness could for the first time be
clearly formulated and even solved. He elaborated Peano's
notation, wrote *The Principles of Mathematics*, and with
Whitehead worked out such matters as the definitions of series,
cardinals and ordinals, and the reduction of arithmetic to
logic. The *Principia Mathematica*, the product of an extraordinary ten-year collaboration, was the crown of Russell and
Whitehead's concerted attack on the complex questions at the
base of mathematics. The complete work appeared in 1913.
The effort was so severe that at the end, as Russell once wrote,
"we both turned aside from mathematical logic with a kind of
nausea." The *Principia* was in part, at least, a magnificent
failure. He had entered upon this mathematical task in search
of "splendid certainty"; yet, as he was later to observe, he had
frustrated himself by finding contradictions not to be reduced

(except by adopting theories "which may not be true and are certainly not beautiful"). And the whole of mathematics, as he has come to believe, though very reluctantly, consists of tautologies, so that "to a mind of sufficient intellectual power" the entire subject would appear trivial, "as trivial as the statement that a four-footed animal is an animal."

"I grew up," writes Russell, "as an ardent believer in optimistic liberalism." His parents had been radicals and freethinkers, friends of John Stuart Mill (who was Russell's godfather). Their will said that their two sons must be brought up as freethinkers. At the request of the grandparents, however, the Court of Chancery set aside the parents' will. The freethinkers whom it designated as the boys' guardians were replaced, and Russell "enjoyed the benefits of a Christian upbringing." But in politics and matters touching on the social good Countess Russell adhered to the family's reforming tradition. At the age of seventy she asserted her independence by becoming a Unitarian. At the same time she supported Home Rule for Ireland, which aristocrats and respectable people regarded as shocking, and was a passionate foe of imperialism and war. Russell shared this outlook. When the *Principia* was finished, he turned to social and political affairs. The coming of the First World War plunged him into self-conflict. He was "tortured by patriotism." "Love of England," he says, "is very nearly the strongest emotion I possess." Yet he thought war was madness and barbarism. Despite his inner turmoil he had no doubt what he must do. He felt as if he had heard the voice of God; he had to protest against war, however futile the protest might be. He collected signatures for neutrality petitions, and after the war broke became intensely active in the pacifist movement. On one occasion, addressing a meeting at a church, he was attacked by a mob. When two drunken ladies began to attack him with boards full of nails, a woman member of his party called on the police to defend him. They merely shrugged their shoulders. " 'But he is an eminent philosopher,' said the woman, and the police still shrugged. 'But he is famous all over the world as a man of learning,' she continued. The police remained unmoved. 'But he is the brother of an Earl,' she finally cried. At this, the police rushed to my assistance."

For more than three years he devoted himself to the cause. Then in 1918 he was sent to prison for pacifist propaganda. The specific charge against him was that he had written a pamphlet accusing the United States Army of "intimidating strikes at home." By the intervention of Arthur Balfour, prison life was made easier for him and he could read and write as much as he liked provided he did no propaganda. In four and a half months he wrote his famous *Introduction to Mathematical Philosophy*, a book that has brought to many of us the first heady experience of Russell's thought and style, and began the work for his *Analysis of Mind*. The governor of the prison found these writings perplexing but unsubversive. Similarly, a warder found nothing objectionable in Russell's reply to a question about his religion. When he said that he was an agnostic, the warder asked how to spell it and remarked with a sigh: "Well, there are many religions, but I suppose they all worship the same God." Russell says "this remark kept me cheerful for about a week."

In September, 1918, he was released. The armistice followed and there was much rejoicing. Russell too rejoiced, but "remained as solitary as before." In 1920 he visited Russia, where he met Lenin, Trotsky and Gorki. He wrote a book, *The Practice and Theory of Bolshevism*, which condemned the Soviet regime for its despotism, its narrow interpretation of Marxist philosophy, its "enormous error . . . in supposing that a good state of affairs can be brought about by a movement of which the motive force is hate." This won him few friends. Conservative opinion condemned him for his views on the war; left-wing opinion scorned him for his betrayal of Utopia.

He spent a year in China, where he was happy. He warmed to the people and found much that was admirable in their tradition. But he feared the effects of "Western and Japanese rapacity" and foresaw the transformation of China into a modern industrial state "as fierce and militaristic as the Powers that it was compelled to resist."

Russell returned to teaching and lecturing, and for several years became absorbed "in parenthood and attendant problems of education." He founded a school that he hoped would promote the best values of "progressive education" by avoiding harshness and deadening disciplinarianism, yet would not be deficient on the purely scholastic side. But he was a poor administrator and the school failed. Nevertheless, his thinking about the matter led to his writing several excellent books on education. In the 1920s and 1930s his literary output was immense. ("Writing books," he says, "is an innocent occupation and it keeps me out of mischief.") He published two dozen volumes and countless journal and general magazine articles on mathematical, philosophical, scientific, political and social subjects. His notable collection of essays, *Mysticism and Logic*, appeared in 1918; *The Prospects of Industrial Civilization*, a study of socialism written in collaboration with his second wife, Dora Russell, in 1923; his well-liked popularization, *The A B C of Relativity*, in 1925; *The Analysis of Matter* and the admirable *Outline of Philosophy* both in 1927; *Marriage and Morals* in 1929; *Education and the Social Order* in 1932; *Freedom and Organization*, a history of political theory, in 1934; and a penetrating analysis of the theory of the state, *Power*, in 1938. Among his later writings are *An Inquiry into Meaning and Truth*, the William James Lectures at Harvard, 1940; *History of Western Philosophy*, 1945, undoubtedly the most palatable survey of its kind published in modern times; and *Human Knowledge*, an epistemological study, in 1948. He has written a book for children, which I remember only

47

vaguely, and a collection of short stories, *Satan in the Suburbs*, which I prefer to forget — though it has not wanted for admirers.

In *My Philosophical Development* Russell sets before us the evolution of his beliefs as to consciousness and experience, language, the nature of truth, the theory of knowledge. In his view we do not directly perceive the external world; what we see are images in our heads from which the world is inferred. But a certain correspondence of structure must be supposed between perceptions and the facts of nature, otherwise understanding would be impossible. This is one of the five principles of "non-demonstrative inference" at which he has arrived. With the whole of Russell's thought synoptically spread out, one is struck by two major tendencies. At times he is a merciless dissector of ideas and language; nothing is accepted at face value, there are no agreed-upons, he cuts familiar notions into thinner and thinner slices as with a microtome. In the *Principia*, for example, it takes a formidable apparatus to prove that one plus one equals two, and he will examine a single word as if it were an infinitely complex crystal, with depths within depths. On the other hand, he will even in a technical philosophical discourse use a plain man's language, now appealing to common sense, now accusing a critic of drowning an issue in subtleties. *An Inquiry into Meaning and Truth* is devoted to demonstrating that, despite the complexity of apparently simple things, and the uncertainty that infects even the most plausible premises, probably the best answer to the question: "How do you know I have two eyes?" is "What a silly question! I can see you have." Russell's easy use of the word "science" (as if its meaning were as univocal and transparent as "Mount Everest"), contrasted with the exhaustive analysis of a word such as "the," must disconcert even his most faithful disciples. His justification for the use of the "common sense metaphysics" on some occasions, and the precision machinery of mathematical logic and analysis on others, is that both are needed to make even a little temporary sense of the world. The philosopher's task is to try to understand the world; but he must recognize that this is impossible, that the most he can hope for is to expose illusion, to demonstrate what is false rather than to establish what is true. Complete skepticism is self-annihilating and insincere; this is where common sense comes in. On the other hand, one must not be too certain of anything, and analysis is needed to show the fugitiveness of timeless truths. Russell's reconciliation of his passion for mathematics and his sympathy for mysticism, of his common sense and scientific outlook, of his reverence for cold and stern perfection and his interest in the less sublime concerns of man, is summed up in these words he once wrote: "I have always ardently desired to find some justification for the emotions inspired by certain things that seemed to stand outside human life and to deserve feelings of awe, . . . the starry heavens, . . . the vastness of the scientific universe, . . . the edifice of impersonal truth which, like that of mathematics, does not merely describe the world that happens to exist. Those who attempt to make a religion of humanism, which recognizes nothing greater than man, do not satisfy my emotions. And yet I am unable to believe that, in the world as known, there is anything I can value outside human beings. . . . Impersonal non-human truth appears to be a delusion. And so my intellect goes with the humanists, though my emotions violently rebel."

For six years, between 1938 and 1944, Russell sojourned in the United States. He lectured at various universities and at the Barnes Foundation at Merion, Pennsylvania. In some quarters he was regarded as what journalists so mischievously call a "controversial" figure. This may be taken to mean he behaved like a man. Also he was guilty of changing his opinions when the evidence seemed to justify change. He had been a pacifist. He had been against the First World War and considered it the cause of such evils as fascism, Communism, Nazism, and the second great war. But he saw no alternative to

49

resisting Hitler's mad ambitions and so, for a few years at least, he was broadly in agreement with his compatriots and condoned war. However, he could never live down in our country some of his rational and unconventional views on society. When he was appointed a professor of philosophy at the College of the City of New York, a lady brought suit to have him barred on the ground that he believed in "free love." A justice of the New York Supreme Court, John McGeehan, sensed the danger at once. He denounced the appointment as an attempt to establish "a chair of indecency." Russell was barred and the students were saved.

We learn about Russell not only through his autobiographical essays but through his sketches of others as well. He has drawn portraits of George Bernard Shaw, H. G. Wells, Joseph Conrad, George Santayana, Alfred North Whitehead, Sidney and Beatrice Webb, D. H. Lawrence. With Shaw, Russell went on a bicycle tour, which had a ludicrous ending. For Wells he felt affection, and admired him as an "important force toward sane and constructive thinking both as regards social systems and as regards personal relations." His contacts with Conrad were infrequent, but from their first meeting the two men were strongly drawn to each other. Conrad, like Russell, was lonely. He was courageous and felt deeply the moral shortcomings of the world, the thinness of civilization's crust. He "despised indiscipline, and hated discipline that was merely external." In all this Russell found himself in close agreement with Conrad. "His intense and passionate nobility," writes Russell, "shines in my memory like a star seen from the bottom of a well. I wish I could make his light shine for others as it shines for me." A kindliness touches Russell's portraits even of men for whom his admiration was not unalloyed. He is not sentimental and he does not fail to recall unpleasing traits, but he has perspective and the charity of wisdom. I am struck by the fact that his set pieces, for all their charm and discernment, describe very little of the subjects' physical appearance or personal habits. Santayana was "prim" and "even in country lanes he wore patent leather boots." But what was he like? Lawrence was full of hatred and jealousy, a cultist with "a mystical philosophy of 'blood.'" But only a part of him comes through. Russell does not always see with the storyteller's eye or the painter's. His appraisals and anecdotes are better than his pictures.

He has not lost his touch, his cutting edge, his capacity for indignation. He can still detect nonsense better than any of us. He can still strike at evil with a formidable set of claws. His eloquence in just causes is as moving as ever. Speaking of man's peril today, he says, "I appeal, as a human being to human beings: remember your humanity and forget the rest." And he grows old so gracefully that one takes pride in the race of men.

"An individual human existence should be like a river — small at first, narrowly contained within its banks, and rushing passionately past boulders and over waterfalls. Gradually the river grows wider, the banks recede, the waters flow more quietly, and in the end, without any visible break, they become merged in the sea, and painlessly lose their individual being. The man who, in old age, can see his life in this way, will not suffer from the fear of death, since the things he cares for will continue. And if, with the decay of vitality, weariness increases, the thought of rest will not be unwelcome. I should wish to die while still at work, knowing that others will carry on what I can no longer do, and content in the thought that what was possible has been done."

LUDWIG
WITTGENSTEIN

D ARKLY wise and a riddle: a strange man and a stranger philosophy, yet each unmistakably touched with genius. Wittgenstein died in 1951, acknowledged as one of the most influential philosophers of the century. His famous work, the *Tractatus Logico-Philosophicus*, appeared soon after the First World War and founded a movement. Two volumes of his papers have been translated and published since his death, and most recently his celebrated *Blue and Brown Books*, dictated in English to his students between 1933 and 1935 and long in circulation in mimeographed form, have been reissued. No full-scale biography has yet appeared, and perhaps none will for a long time, but the Malcolm memoir* provides an intriguing entry into the life of this

---

* Norman Malcolm, *Ludwig Wittgenstein, a Memoir*, Oxford, 1958.

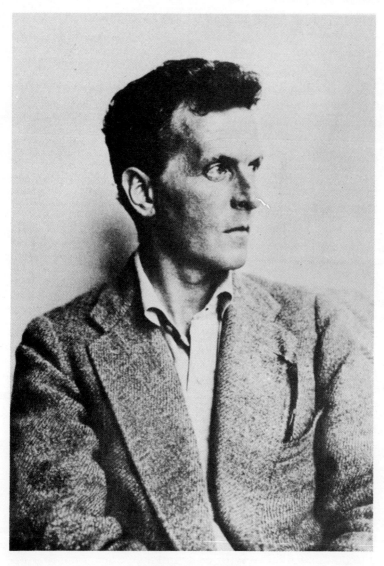

Ludwig Wittgenstein.
(From Dagobert Runes, Pictorial History of Philosophy,
by permission of Philosophical Library)

extraordinary thinker, and its interest and value are enhanced by an excellent prefatory biographical sketch written by Georg Henrick von Wright, professor of philosophy at the University of Helsingfors.

Ludwig Joseph Johann Wittgenstein was born in Vienna in 1889. The family, which had migrated from Saxony to Austria, was of Jewish descent, but his grandfather had been converted to Protestantism and Ludwig Wittgenstein was baptized a Catholic, the religion of his mother. It was a wealthy and cultured home. His father was an engineer prominent in the steel and iron industry, and both parents were artistically and intellectually inclined, with a deep interest in music. Johannes Brahms was a close friend of the family.

Ludwig was the youngest of five brothers and three sisters. Deep instability as well as talent marked the children: three of his brothers were later to commit suicide. He himself lived always "on the border of mental illness," fearful of being driven across it; but it would be wrong, says Von Wright, to think of him as morbid. He was strikingly original, often difficult to follow, and certainly given to eccentricity; his later work suffered more and more from an almost wayward obscurity, but his earlier labors had, in Von Wright's view, "the same naturalness, frankness and freedom from all artificiality that was characteristic of him."

Until he was fourteen, Wittgenstein was educated at home. Thereafter he attended school in Linz in Upper Austria and a technical high school in Berlin. At first he intended to study physics under Boltzmann in Vienna, but the latter's death put an end to this wish and Wittgenstein turned to engineering. (Machines continued to fascinate him throughout his life. Even in his last years he could spend a whole day with "his beloved steam engine" in the South Kensington Museum, and there are anecdotes about his serving as a mechanic when some contraption got out of order.) In 1908 he went to England and registered as a research student in engineering at the University of

54

Manchester. For three years he focused his attention on aeronautics — studying jet engines, experimenting with kites, working on propeller design. The mathematical aspect of propeller design absorbed him, and by that route he came to pure mathematics and the foundations of the subject. He had been unhappy and restless in his engineering researches, but in the new world of abstractions he found deep and growing satisfaction. He read Bertrand Russell's *Principles of Mathematics*, which was a revelation, and Gottlob Frege's writings on symbolic logic: this was the gateway through which he entered philosophy. On the advice of Frege, whom he visited in Jena, he entered Cambridge to study with Russell.

During 1912-13 he was at Trinity College, working as an advanced student in philosophy and psychology. This was a great period in the intellectual life of the university; Wittgenstein thrived on the rich ferment of ideas. He became friendly with John Maynard Keynes, G. H. Hardy and the logician W. E. Johnson; he was very close to David Pinsent, a young mathematician who fell in the war (the *Tractatus* is dedicated to his memory); he saw much of G. E. Moore and Alfred North Whitehead. With Bertrand Russell he formed an intimate relationship, and each man was to have a profound influence upon the thought of the other. At first Russell had difficulty in appraising this queer fish. Was he a genius or merely an eccentric? At the end of Wittgenstein's first term at Cambridge he came to Russell (who tells the story) and said: " 'Will you please tell me whether I am a complete idiot or not?' I replied, 'My dear fellow, I don't know. Why are you asking me?' He said, 'Because, if I am a complete idiot, I shall become an aeronaut; but if not, I shall become a philosopher.' I told him to write me something during the vacation and I would then tell him whether he was a complete idiot or not. At the beginning of the following term he brought me the fulfillment of this suggestion. After reading only one sentence, I said to him, 'No, you must not become an aeronaut.' " They took long walks together; Wittgenstein would frequently come to Russell's rooms at midnight and for hours "would walk backward and forward like a caged tiger." Wittgenstein would announce that when he left he would commit suicide. "On one such evening, after an hour or two of dead silence, I said to him [Russell writes], 'Wittgenstein, are you thinking about logic or about your sins?' 'Both,' he said, and then reverted to silence." He must have been trying, but enormously stimulating too. "Getting to know Wittgenstein," Russell said in a memorial article, "was one of the most exciting intellectual adventures of my life."

Wittgenstein entered the Austrian Army as a volunteer at the outbreak of war. At first he fought on the eastern front, but in 1918 he was transferred to the south; when the Austro-Hungarian army surrendered, he was taken prisoner by the Italians. Men of passionate intellectual bent take their work with them everywhere, and Wittgenstein, like Descartes, did not find soldiering an insuperable obstacle to creative thinking. When he was captured he had in his rucksack the manuscript of *Tractatus*, which he had completed when on a leave of absence in Vienna in August, 1918. He had thought about its logical questions before the war and had formed certain of his major conclusions, but the idea of language as a picture of reality occurred to him, as he told Von Wright, one day in a trench on the eastern front while he was reading a magazine in which there was a picture of the possible sequence of events in an automobile accident. The picture, he said, served as a proposition whose parts corresponded to things in reality; and so he conceived the idea that a verbal proposition is in effect a picture, "by virtue of a similar correspondence between *its* parts and the world." In other words, the *structure* of the proposition "depicts a possible combination of elements in reality, a possible state of affairs." The *Tractatus*, as Von Wright observes, may be called a synthesis of theory-of-truth functions

56

and the idea that language is a picture of reality, giving rise to the doctrine "of that which cannot be *said,* only *shown.*"

While still a prisoner in Monte Cassino Wittgenstein was able, through the intervention of his friend Keynes, to send the manuscript of the *Tractatus* to Russell and a copy to Frege. Before the war Wittgenstein had given Russell a short type-script consisting of notes on various logical points; these, together with what Russell had gained from their many con-versations, strongly affected his thinking. The manuscript of the *Tractatus* stirred him even more profoundly. Writing in a British magazine the early part of 1959, Russell said that "I do not feel sure that either then [1914] or later, the views which I believed myself to have derived from him [Wittgen-stein] were in fact his views"; adding that Wittgenstein always "vehemently repudiated expositions of his doctrines by others, even when those others were ardent disciples." Yet whether or not the *Tractatus,* at times a distressingly aphoristic and Del-phic piece of writing, successfully communicated its author's meaning, or whether its influence (as Russell now questions) was "wholly good," there can be no doubt of its dominating effect on the British philosophical world. It cut a new channel for twentieth-century thought.

Though he was able, despite the mortal uncertainties and confusion of the time, to complete the *Tractatus,* it cannot be imagined that the war gave Wittgenstein ease of mind. He became acquainted during this period with the ethical and re-ligious writings of Tolstoy, which had a strong influence on his view of life and also led him to study the Gospels. After his release as a war prisoner, Wittgenstein entered a Viennese college for elementary-school teachers. He took a two-year course, then for six years served as a schoolmaster in various remote Austrian villages. He had no private means, because, though his father had in 1912 left him a great fortune, imme-diately after the war he gave all his money away and practiced

57

a Tolstoyan simplicity and frugality.* His secluded life as a schoolteacher suited his mood but not his temperament. He was in "constant friction" with the people around him, and finally, after a serious crisis, resigned his post. For a time he worked as a gardener's assistant in a monastery near Vienna and even considered becoming a monk. In the autumn of 1926 he tried his hand at architecture. He built a mansion in Vienna for one of his sisters. Von Wright describes it as a work "highly characteristic of its creator down to the smallest detail ... free from all decoration and marked by a severe exactitude in measure and proportion." The materials were concrete, glass and steel; it had horizontal roofs and its beauty was "of the same simple and static kind that belongs to the sentences of the *Tractatus*."† He also did sculpture but along more classical lines than he followed in architecture; one of his creations is "the head of a girl or an elf."

It was evident that he was restless, troubled, uncertain of his course. He had written his book, it said all he had to say, and he saw no reason to continue in philosophy. "I am of the opinion," he wrote in the preface to the *Tractatus*, "that the problems have in essentials been fully solved." He had turned philosopher, as John Passmore remarks, "in his engineer's way, in order to drain what seemed to him a swamp. The task was completed; there was no more to be said." However, the actual publication of the *Tractatus* in 1922 had not been a joy either to him or to Russell. The circumstances were typical. Russell went to Holland to discuss the manuscript with him. It was hard to find a publisher, but Russell succeeded and then wrote an introduction that Wittgenstein strongly disapproved. Finally he turned his back on the whole undertaking.

---

* Since writing this I have been told by the mathematician Karl Menger, who knew Wittgenstein and members of his family, that while it is true he gave his money away, the gift was to his sisters, on the understanding that he could always draw on them for what he needed. He simply wished to be disencumbered.

† Frank Lloyd Wright, I am told, saw the mansion and greatly admired it.

But of course he was not through with philosophy. While he was still a schoolteacher, the brilliant young philosopher Frank Ramsay, who had assisted in the translation of the *Tractatus* and had written an admirable review of it for *Mind*, sought out Wittgenstein in Austria and tried to persuade him to return to England. These conversations with Ramsay, Wittgenstein says, woke him from his dogmatic slumber. With the help of Keynes, who raised the money, Wittgenstein visited his English friends in 1925. Another of his philosophical contacts during this period was the Austrian Moritz Schlick (who was to become founder and leader of the Vienna Circle), who had seen the *Tractatus* and esteemed it a first-class work.

In 1928 Wittgenstein heard the Dutch mathematician L. E. J. Brouwer, founder of the intuitionist school of mathematical philosophy, lecture in Vienna on the foundations of mathematics. This was the spark that revived Wittgenstein's interest in philosophy. He now felt that he could again do creative work. In 1929 he returned to Cambridge as a research student and the following year was elected a fellow of Trinity College. During the next three or four years he wrote a good deal on the philosophy of mathematics, but published only a single article (on logical form). Those who had access to his manuscripts were impressed with his ideas, but he was evidently in a transitional stage of development and "fighting his way out of the *Tractatus*." The break with the past occurred about 1933. "There came to him at this time," says Von Wright, "those basic ideas whose development and clarification absorbed him for the rest of his life." Frege and Russell had contributed much to the shaping of the ideas in the *Tractatus*; their views were undoubtedly Wittgenstein's point of departure, but the philosophical ancestry of his later views, of which so many echoes can be heard in the writings of the Vienna Circle, is less clear. It has been said that the influence of G. E. Moore is plainly visible, but Von Wright dismisses this suggestion, arguing instead that the "impression of resemblance" between the views of both men is due to the influence both had on the trend known as analytic or linguistic philosophy. Norman Campbell's *Physics* and William James's *Principles of Psychology* — for a time James's work was the only one visible on Wittgenstein's bookshelf — have been mentioned as playing a part in the growth of his "new" philosophy. In any case he now came to repudiate some of the fundamental thoughts of the *Tractatus*, thoughts that had been strongly criticized by Ramsay, and by another of Wittgenstein's friends, the Italian economist Piero Sraffa. Wittgenstein said that his discussions with Sraffa, when he taught at Cambridge, "made him feel like a tree from which all branches had been cut."

For the rest of his life, with some interruptions, Wittgenstein lived in England, teaching for many years at Cambridge. When the Germans swallowed Austria he became a British subject, but for all the years he lived in England, for all the freedom and security and intellectual nourishment it gave him, it never became his country. He could not get used to English ways of life and he disliked the academic atmosphere of Cambridge. In 1935, when his Trinity fellowship expired, he planned to settle in the Soviet Union, and a visit there with a friend pleased him. Events of the middle 1930s, however, contributed to the abandonment of his plan. For a year (1936) he lived in solitude in a hut he owned in Norway, where he began to write the *Philosophical Investigations*. In 1937 he returned to Cambridge and two years later succeeded to Moore's chair in philosophy.

Malcolm, an American student holding a Harvard fellowship at Cambridge, met Wittgenstein in 1938. The memoir gives a vivid impression of him at that time. Though he was forty-nine years old, he looked about thirty-five. He was some five feet six inches in height, slender, with curly brown hair. His face was lean and brown, his profile aquiline and "strikingly beautiful," and he had deep eyes, which were often "fierce" in expression. The effect was of a commanding, even "imperial," personality, which instantly drew attention at any gathering. Sometimes he was hesitant and would stammer, but usually he was emphatic and expressive, though obviously deeply immersed in his thoughts and desperately anxious, as one would expect from the bone and blood of his philosophical creed, to convey his exact meaning. He spoke excellent English, "although occasional Germanisms would appear in his constructions."

In contrast to his manner, his dress was extremely simple. "He always wore light grey flannel trousers, a flannel shirt open at the throat, a woolen lumber jacket or a leather jacket. Out of doors in wet weather, he wore a tweed cap and a tan raincoat. He nearly always walked with a light cane. One could not imagine Wittgenstein in a suit, necktie or hat. His clothes, except the raincoat, were always extremely clean and his shoes polished."

Twice a week he met his class for two hours, from five to seven. Students brought their own folding chairs, and if someone came a little late this involved a disruption, because chairs already placed in the crowded room had to be moved. Tardiness made Wittgenstein very angry; the latecomer had to face his formidable glare. His rooms were austerely furnished. There was no easy chair or reading lamp. The walls were bare. He slept on a canvas cot, and in the living room were two canvas chairs and a plain wooden chair, with an old-fashioned iron stove to provide heat. The only adornments were flowers in a window box and in a couple of pots. He did his writing on a card table and kept his manuscripts in a metal safe.

During class Wittgenstein sat on a plain wooden chair in the center of the room. "Here he carried on a visible struggle with his thoughts." This often led him to feel that he was confused; he would say "I am a fool" or "You have a dreadful teacher." He thought about certain problems as he went along; his classes were more conversation than lectures. When he was

61

wrestling with an idea, he would "with a peremptory motion of the hand" prohibit further discussion, so that there were frequent and prolonged periods of silence "with only an occasional mutter from Wittgenstein, and the stillest attention from the others."

It is easy to imagine that he frightened his students. He was impatient, irritable, insulting. He drove himself, scourged himself to achieve understanding, and was no more sparing of his class. They respected his passionate honesty and, if not petrified, were spurred to hard mental exertion; but two hours of such severe and tense probing exhausted both teacher and pupils. When it was over, Wittgenstein, full of self-reproach, would eagerly seek relief from his ordeal. Often he would "rush off to a cinema immediately after the class ended," taking a friend with him. He would buy a bun or an execrable English cold pork pie and munch it while he watched the film. He insisted on sitting in the front row so that the screen "would occupy his entire field of vision, and his mind would be turned away from the thoughts of the lecture and his feelings of revulsion." No matter how wretched the picture, he became totally absorbed in it. He liked United States films and detested British ones — part of his distaste for English culture; Carmen Miranda and Betty Hutton were two of his favorite actresses.

It was important to Wittgenstein to make friends with members of his classes. Philosophy went better, he felt, with friends, and he was stimulated by the sight of "friendly faces." He liked to make jokes to illustrate a point, and to laugh at his own cleverness, but if a member of the class laughed, Wittgenstein would reprove him. Facetiousness had no place in philosophical discussions, at least the facetiousness of others. He was a trying man.

He would invite students individually for tea, but no small talk was permitted. As in class the conversation was interspersed with long silences. Though a teacher of philosophy, he would try to persuade students to give up the subject. This was not so much because he felt they had no useful contribution to make as that he hated academic life in general and professional philosophers in particular. A normal human being, he believed, could not be a university teacher and also an honest and a serious person. He could not stand the society of academic colleagues; he would not dine in the Hall of his college, "being revolted by the artificiality of the conversation."

Wittgenstein argued with Malcolm to abandon philosophy and take a manual job. When it became clear that Malcolm intended perversely to follow the less honest but more mental career, Wittgenstein gave him money so that he could continue his studies at Cambridge. They became good friends and saw much of each other. They would take long walks that were apt to be exhausting. The tempo was in fits and starts. Wittgenstein would walk rapidly in spurts, would stop suddenly to look into Malcolm's eyes with a "piercing gaze" and then would trot off again, all the while expounding the most serious and even gloomy thoughts, erratically punctuated with "jests," which he alone was permitted to laugh at. He was exquisitely touchy. Once on a walk in 1939 Wittgenstein and Malcolm saw a news vendor's sign which announced that the German government accused the British government of instigating an attempt to assassinate Hitler. Wittgenstein suggested the German claim might be true; Malcolm repudiated the idea on the grounds that the British were "too decent" to attempt anything so underhanded, and that it was against their "national character." Wittgenstein became extremely angry, told Malcolm that he was stupid and that he had evidently learned nothing from the philosophical training Wittgenstein was trying to give him. When Malcolm refused to back down, Wittgenstein would not talk to him any more; soon after, they parted. Wittgenstein came no more for his walks, and although the friendship was resumed later, the episode was never forgotten. Wittgenstein managed to be enraged even by G. E. Moore.

Having read one of Moore's papers with which he disagreed, he went to Moore's home and for two hours furiously harangued him, giving him no chance to reply. Later, when Moore told him he had been rude, he made a "stiff and reluctant apology."

In February 1940 Malcolm returned to the United States. He had been ill just before his departure, and Wittgenstein came to see him as a gesture of reconciliation. He fussed over the patient, and said he was sorry about the assassination argument. He also gave the sound advice: "Whatever else you do, I hope that you won't marry a lady philosopher!" The two men kept up a correspondence. Wittgenstein was an avid reader of detective magazines, and as a connoisseur preferred those of Street and Smith; these being unobtainable in England during the war, Malcolm would send him bundles of them from time to time. Wittgenstein said he could not understand why anyone would read *Mind*, "with all its impotence and bankruptcy, when they could read Street and Smith magazines." In at least one case a story so entranced him that he wanted to write the author and thank him. "If," he wrote Malcolm, "this is nuts, don't be surprised, for so am I." When Malcolm got his Ph.D., Wittgenstein congratulated him and wrote: "I wish you good, not necessarily clever, thoughts, and decency that won't come out in the wash." On another occasion Wittgenstein wrote to Malcolm: "I wish you could live quietly, in a sense, and be in a position to be kind and *understanding* to all sorts of human beings who *need it*! Because we all need this sort of thing very badly."

For part of the war Wittgenstein worked as an orderly in Guy's Hospital in London; later he worked in an infirmary and in a clinical research laboratory. He could not do philosophy and he was "tired and sad." In 1944 he was back at Trinity, writing *Philosophical Investigations*. He thought it was "pretty lousy" but felt he couldn't improve on it if he tried "for another 100 years." He was reading Johnson's life of Pope in *Lives of the Most Eminent English Poets*, also his *Prayers and Meditations*, which he liked very much. He commented that Freud was both "charming" and "full of fishy thinking," but that this did not detract from his "extraordinary scientific achievement." In 1946 Wittgenstein's lectures went pretty well, but at the end of the term he said he felt his brain was "burnt out, as though only the four walls were left standing, and some charred remains." The same year Malcolm returned to Cambridge, which gave Wittgenstein the opportunity to meet his wife. At first he was suspicious of her, as he was of all "don's wives," but he got over this and became a frequent visitor. After supper he often insisted on washing the dishes, a feat performed in characteristic style with great thoroughness in the bathtub.

There are many more anecdotes in Malcolm's memoir. They reinforce the impression of Wittgenstein's complex and difficult personality. He was terribly hard on his friends, and just as hard on himself. His suspicions and fears, amounting almost to mania, kept disrupting his relationships. But people put up with him not alone because of his extraordinary gifts, but also because of his ruthless honesty and sincerity. When he was charming or gracious or kind — as he could be — his purity shone through.

In the fall of 1947, after a trip to Austria, he resigned his professorship. He was not finished with philosophy, but he had reached the end of his tether in the academic world. He spent some time in Eire. He was depressed, ill in body and mind. In a lonely cottage on the western coast right on the sea, seeing nobody except the man who brought his milk, he dreamed, brooded, walked, watched the sea birds, worked a little. He didn't miss conversations, he said, but "only someone to smile at occasionally." Malcolm's shipment of detective magazines put him in better spirits. The concern of the Malcolms for this sad, proud, fading man was touching. They invited him for a long visit to Cornell. He envisaged all sorts of difficulties, but

65

he was eager to come and the matter was arranged. Malcolm met him when the Queen Mary docked. He came "striding down the ramp with a pack on his back, a heavy suitcase in one hand, cane in the other." He seemed vigorous and in excellent spirits.

At Ithaca he was nearly his old self, taking long walks, worrying linguistic problems, indulging in small tyrannies and eccentricities. Even his diversions were intellectual. For example, on one walk he decided to measure the heights of the trees. The procedure was that he would place himself at a distance from the tree, sight along his arm and cane at the top of the tree, his arm at about a forty-five-degree angle from the horizontal; then Malcolm would pace off the distance to the foot of the tree and Pythagoras would provide the approximate height. Wittgenstein directed this activity with "real zest." Malcolm's wife once gave Wittgenstein some Swiss cheese and rye bread for lunch, which he greatly liked. Thereafter he asked for this combination at all meals, "largely ignoring the various dishes that Mrs. Malcolm had prepared." He explained this by saying "it did not much matter to him *what* he ate, so long as it was *always the same.*"

Wittgenstein met with Malcolm's colleagues and graduate students to discuss philosophy. At these sessions there were signs of his old fire, but illness gradually encroached upon him and he had to forego attendance. He was not too ill to be insulting. When he suggested that the *Philosophical Investigations* might be mimeographed, and Malcolm disparaged the idea, Wittgenstein was angered and accused him of being reluctant to see the work made public "because people would then know where my own [Malcolm's] philosophical ideas came from." Malcolm had merely meant that it was not fitting that a book of such importance should be distributed in mimeograph: "It should be bound in 'leather and gold.'" One more anecdote catches the great logician at work. It was a hot summer, and Wittgenstein's room at the Malcolms' was often stuffy. He suggested removing the window screens to permit freer circulation of the air. Malcolm pointed out that the insects would be worse than the heat. Wittgenstein doubted this, and referred to the fact that screens were rare in England and on the Continent. Malcolm answered that we had more insects, which Wittgenstein didn't believe. He went out for a walk to see whether other houses had screens or whether this was simply one of Malcolm's peculiarities. On finding that all other windows had screens, he inferred not that there must be a good reason for it, but ("with some irritation") that "Americans were the victims of widespread and unthinking prejudice as to the necessity of window screens."

In October he returned to England. The latter part of his stay had been wretched because of illness. He was sure he had cancer, and while he was quite prepared to die, he was afraid he would be kept at the hospital for surgery. He didn't want to die in America. "I am a European — I want to die in Europe," he said to Malcolm in a frenzy. He had his wish. In December he wrote the Malcolms from Cambridge that he had cancer of the prostate. Medication alleviated the symptoms of the disease; he was able to go about, even to visit Vienna. He read various odds and ends and was not depressed. For a few months he lived in Oxford; then he went to Norway for a month. In 1951 he returned to Cambridge to stay in the home of his physician. He even resumed work. He died very suddenly in April. Though much of his life had been spent in unhappiness, even torment, his last words to the wife of his host were "Tell them I've had a wonderful life."

THE AGE
OF ANALYSIS

M ORTON WHITE'S *Age of Analysis** is the
most recent addition to a well-thought-
out series whose purpose is to acquaint the general reader with
the ideas of the leading philosophers of the Western world by
giving excerpts from their writings, interlarded with interpre-
tive commentary. Six volumes will span philosophy from
medieval through present times; Professor White's survey,
though chronologically the last, is the second to be published.

I should make it clear at the outset that this is not a book
for hammock reading at Old Point Comfort. Philosophy is not
a cheap and easy enterprise, though there are popularizers

---

* *The Age of Analysis*, edited by Morton White, Boston, 1955.

who have cheapened it; the philosophy of the age of analysis, concerned with logic, linguistics and science, can be as disagreeably difficult as any branch of thought yet devised. One of the merits of this survey is that it does not attempt to pretty up the subject. Most popular anthologies concentrate on the philosophers who make agreeable, sympathetic, charming or elevating reading. But because it is fair to assume that there is an audience anxious to hear what "the other half" of philosophy says, White has included passages from the hard-bitten, technically minded analysts as well as from the bold and mellifluous system builders.

White suggests dividing the philosophers of the twentieth century into two groups, hedgehogs and foxes. The image (recently revived by Sir Isaiah Berlin in an essay on Tolstoy) is from a line of the Greek poet Archilochus which says: "The fox knows many things, but the hedgehog knows one big thing." We must not expect too much of this image, but it serves to contrast two main tendencies. On one side are the metaphysicists, the builders of monuments, who try to see the world in terms of a central concept which is to organize all their attitudes and beliefs. These are the hedgehogs. In their systems there is a place for everything and everyone; all that the seen and unseen world has been, is, and will be is explained. Their philosophies may not always be a comfort, but they are always complete. There are no loose ends or exceptions; loneliness and starfish, molecules and bills of lading, greed and galaxies, time and art are embraced in a single vision. The other tendency is both more modest and more arrogant. Its followers have a poor opinion of metaphysics. They deny that philosophy has to do with world views, that it has any business pronouncing grandly on religion, politics, morals, art or even science. Their tradition offers little support either to political or religious movements, which is one of the reasons why it has so many detractors. The logical-analysis foxes would like to know many little things — instead of one

69

big thing — but they are content to know even *one* thing, provided they can get to know it very well. One may vary the image of the foxes, and think of this second tendency as surgical. Using the sharpest possible instruments of logic and mathematics, these surgeons of philosophy are intent on excising small muddles that, if untended, may grow into big muddles. The object of attention is not the world or man or morals, but sequences of reasoning, sentences, even single words. Logical analysts are philosophers interested in the causes and cure of philosophy. In time they hope to put themselves out of business.

This, then, may be regarded as the main cleavage of twentieth-century philosophy. I have, to be sure, oversimplified it. There are system builders who focus on details and analysts who seek to put together small things into a bigger thing. Moreover, between the hedgehogs and the foxes there is a middle species that aims to bridge the gap — notably the Americans Charles Peirce, William James and John Dewey. In any case, there are both a gap and a great variety among the doctrines and concepts of both sides, which is what makes modern philosophic thought so engrossing.

Among the great system builders are Plato, Aristotle, Descartes, Locke, Hume, Kant, and, more recently, Hegel. Hegel is of particular interest because it is this "enormously muddled but brilliant German professor of the nineteenth century" who most profoundly affected the philosophy of the twentieth century. Today he has few disciples, but many offspring — most of whom deny him. One must look to his repudiators to judge his importance. As White points out, not only did Hegel "influence the originators of Marxism, existentialism and instrumentalism — now three of the most popular philosophies in the world — but at one time or another he dominated the founders of the more technical movements, logical positivism, realism, and analytic philosophy." Hegel regarded the uni-

Hegel.
(The Bettmann Archive)

71

verse as the unfolding of a "World Spirit" or "Absolute." The Absolute is spiritual, and resembles an animate entity in having direction, desires, purposes and the like. Human actions, historical changes and institutions are imbedded in this cosmic organism and are to be understood as creatures of its working. The process by which the Absolute manifests its will is called the "dialectic." This is a kind of triadic ballet of thesis, antithesis, and synthesis, ruled by a special sort of logic that controls the pattern of change and development. By this curious mode of locomotion, like that of a sea serpent making and dissolving its great loops as it advances, history slithers on, knowledge grows, the Absolute journeys toward its fulfillment. It is unnecessary to tarry in this Absolute thicket, but there are several points to keep in mind about Hegel. Many thinkers besides himself professed to understand his doctrines; he shaped the course of modern history as well as philosophy; his system provided a scaffolding for religious belief, thereby attracting those "who could not accept atheism or Kant's peculiar agnosticism"; he encouraged the feeling, which today reigns in science as well as other branches of thought, that "there are modes of explanation other than those available in Newtonian mechanics." He straightened out some things, but he turned many more things upside down. Thus he made work for the philosophers who came after him. So strong was his appeal — he captured in their youth such men as Dewey, Bertrand Russell and G. E. Moore — that philosophers had, it seemed, to go through their Hegel period (as young literary men had at one time to go through their Werther or Byron period) before they could work through to their own systems, and devote themselves to the antithesis of their first thesis, a process called dehegelization.

The hedgehogs represented in this book are Benedetto Croce, George Santayana, Henri Bergson, Alfred North Whitehead, Edmund Husserl and Jean-Paul Sartre. Croce, who died in 1952 at the age of eighty-six, was Italy's most distinguished philosopher. Wealthy from birth, he turned first to the law, then to antiquarian studies and philology, finally to philosophy, literature and history. Closer to Hegel than any other thinker discussed in these pages, Croce was an idealist. This is not easy to define beyond saying that idealists assign to ideas the highest kind of existence. What is real is what is in our heads; buttonhooks and chairs are in a doubtful category because all we know about them, their entire existence, is demonstrated only in mental activity. Two elements enter into our judgment about anything: an immediate awareness or experience, and a linking of this experience with others like it to form abstractions or concepts. This is the sum total of our knowledge; if these elements are removed there is nothing, and the supposed world of substance vanishes. In his principal work, *Philosophy of the Spirit*, he applied his doctrines to aesthetics, logic, economics, ethics and history. His most striking conclusion is that since history "is the concrete study of the spirit, of life, of human activity" it is identical with philosophy itself.

The cult of history, which would not have been altogether displeasing to Hegel, provoked a strong reaction in Santayana, the Spaniard who for many years taught at Harvard and then became an expatriate. In his outlook moral philosophy occupied the central position and history was no more than a "servile science." The main theme of his five-volume *The Life of Reason* is the "transformation of man's natural impulses into high ideals." A survey of man's striving shows him trying to *live* the life of reason, which reaches its highest expression in religion, art and science. Santayana was a brilliant writer. The suavity and elegance of his prose are in fact a little benumbing. One is apt to agree with what he says without knowing exactly with what one is agreeing. Bertrand Russell compared Santayana's style to the patent-leather boots he wore: too smooth and polished. "The impression one gets in reading him is that of floating down a smooth-flowing river, so broad that

73

Santayana during his professorship at Harvard University.
(The Bettmann Archive)

you can seldom see either bank; but when, from time to time, a promontory comes into view, you are surprised that it is a new one as you have been unconscious of movement."

Another leading figure who produced a similar hypnosis-by-eloquence was Henri Bergson. No more, however, than Santayana can he be flippantly dismissed. Bergson was as important for what he said that was foolish as for what he said that was sensible. He was an "outspoken irrationalist," a relentless foe of "scientism," positivism and materialism. This explains his immense popularity among artists, writers, religious thinkers and others who feared that mechanistic philosophies would destroy the values they cherished. It also explains the resurgence today of similar anti-intellectual tendencies. Contemporary anti-intellectualism is by no means confined to mischievous politicians. Biologists are busy locating spirit in organism, physicists see God in the harmony of certain natural constants, and psychiatrists get into bed with German theologians. The key words of Bergson's philosophy are *"élan vital"* (the vital impulse, more fundamental than either mind or matter), "enduring" and "creative evolution." The intellect, he felt, was good in its way but not in all weathers. Science has its uses but because its progress depends on effacing individuality and qualities, it is dependable only when it deals with the inert. When it undertakes the study of life, it furnishes, at best, pale, geometric images from which the uniqueness and originality of life have been drained. Life is to be lived, not merely thought about or known; the heart as well as the mind must play a part. In enduring we prove and fulfill ourselves; in elevating intuition and instinct over intellect we penetrate to the *"real* reality." Evolution is creative precisely because it is more than mechanical, because there is room in it for "real change and real freedom, unhampered by causality and determinism." The crowds, including the "fashionable ladies," who flocked to hear Bergson's lectures at the

Henri Bergson.
(The Bettmann Archive)

Collège de France were enthralled, and William James greeted the appearance of *Creative Evolution* with "ecstasy." Russell was less thrilled. Anyone, he said, looking for evidence to support such a restless view of the world "will find, if I am not mistaken, there there is no reason whatever for accepting this view, either in the universe or in the writings of M. Bergson."

Alfred North Whitehead belongs in the company of Croce and Bergson, though as White points out, this grouping would not have been foreseeable thirty years ago. Until then he was known for his work in mathematics and logic, his stupendous collaboration with Russell on the *Principia Mathematica*, and his profound studies in the philosophy of science. But in the mid-twenties he crossed over to the metaphysical side, repudiated the tradition he had done so much to further, and turned his thoughts to religion, education, morals, history and the "dark and difficult notions of intuition and organism." Once an apostle of clarity and precision, he came, after his conversion, to scoff at them. One of his basic ideas is that nature is alive. Bergson was content to say that a *part* of nature — the most important part — is alive and that a true philosophy has to recognize the fact. Whitehead went further. He deemed it absurd to regard *any* part as dead: "All ultimate reasons are in terms of aim at value. A dead Nature aims at nothing."

Many acute thinkers, having given Whitehead's system careful and respectful attention, have despaired of understanding it. It is easier to explain what he was *against* than to expound the positive features of his system. "Basically," writes White, "it is the view that nature is composed of permanent things, bits of matter moving about, as he says, in a space otherwise empty, each one having its shape, its mass, its motion, its color, its smell. It is the view of the great thinkers of the sixteenth and seventeenth centuries, the view of the ordinary man today, and according to Whitehead it has never been successfully extruded from the minds of scientists even though science has

77

Alfred North Whitehead. Pencil portrait by Paul Drury, collection of Trinity College, Cambridge. (Photo: Stearn and Sons, reproduced by permission of Trinity College, Cambridge)

thoroughly discredited it." Modern science has made it clear that there is no empty space, but only fields of force; that neat packets of matter are a fiction; that matter is energy and energy "incessant activity." Everything therefore is an "event." All events are intertwined, and when a leaf trembles, the universe shakes. There are echoes of Bergson in Whitehead's universe of process. It is neither unintelligible nor merely a machine. It is a universe of becoming and perishing, of "creative advance." It is not necessary fully to grasp Whitehead's meaning to realize that his system transforms profound scientific insights into a comprehensive philosophy of man and spirit.

I pass quickly over the German phenomenologist Husserl and the distinguished French literary man, critic and philosopher Jean-Paul Sartre. Husserl's ideas are to me peculiarly opaque. Sartre's existentialism (which owes much to the Danish theologian Sören Kierkegaard, and is anticipated in the writings of Dostoevski, Nietzsche and Franz Kafka) appears in both a moderate and an extreme form. One can be a Catholic existentialist or, like Sartre, an existential atheist, or somewhere between. Existentialism is essentially an anti-intellectual outlook, and peculiarly appealing to those who see ethical ideals lying in ruins and seek solace in a categorical imperative of absolute personal responsibility for conduct. By my imperfect grasp, existentialism is more of a creed than a full-grown philosophy.

Of the famous American trio of pragmatists, Peirce, James and Dewey, the first was the philosopher of science, the second the philosopher of religion, the third the philosopher of morals. James once made a celebrated division of philosophers into the "tender-minded" and the "tough-minded." Peirce was tough-minded. He was original, wayward, prickly, unsuccessful — "a brilliant unemployable who had to be befriended by saintly people like William James." He introduced into philosophy the word "pragmatic," which comes from a Greek word meaning "action" and is the source of our words "practice" and "practical." The main task of pragmatic philosophy he conceived to be the clarification of the meanings of words — nouns and adjectives — as they are used by scientists. Take, for example, the word "hard." To specify its meaning and make sure it always means the same thing, we must define it by translating the sentence "This is hard" into something like "If one were to try to scratch this, one would not succeed." In general, all descriptive statements in science should be translatable into an if-then form: "If operation O were to be performed on this, then E would be experienced." The entire procedure has a hypothetical, an operational and an experiential element. Any term, however well established, that resists this approach, is meaningless. If any two terms yield the same translation, they are the same, however different they may appear. Peirce said that "the whole function of thought is to produce habits of action." If we respond to a word in the way Pavlov's dogs responded to a dinner bell, the word is doing its job; otherwise not. One of his famous rules reads: "Consider what effects, which might conceivably have practical bearings, we conceive the object of our conception to have. Then, our conception of these effects is the whole of our conception of the object."

James was tenderer than Peirce. He had had things easier. Also he was a fuzzier thinker, but I think deliberately, just as the later Whitehead preferred fuzziness. He wrote a great work on psychology, was interested in the individual and his problems, and reflected deeply on the foundations of belief and the pragmatic theory of truth. For James there was no finality in truth. Truth is concrete, adequate, useful, empirical. It changes as we change. He quotes the German chemist Ostwald reporting on a wrangle between chemists over the word "tautomerous." The argument, said Ostwald, would never have begun if any one of the parties to it had ever asked himself what experimental fact "could have been made different by one or the other view being correct." The dispute would then have been exposed as having as little substance as a dispute among primitive men over whether dough rises because of the exertions of a "brownie" or an "elf." Since truth is to be measured not alone by correspondence but also by consequences, James suggests the maxim that the truth is "what we ought to believe"; and what we ought to believe is "what is better for us to believe." Admittedly it is not always easy to discover what is "better for us." Benefits have to be weighed, and the best we can hope to arrive at is "probable truth." But this is to be tested not by principle or dogma or "skinny" abstractions but rather by "what works best in the way of leading us, what fits every part of life best and combines with the collectivity of experience's demands, nothing being omitted."

John Dewey detected in James's test of truth what White characterizes as a certain "capriciousness." His philosophy, called instrumentalism or experimentalism, reverts to the "social and public pragmatism" of Peirce, and is more concerned with social or public problems than Peirce's or James's philosophy. His ethical theory, he hoped, would mediate between the ethics of "transcendental eternal values" and the view that value is determined "by mere liking, desire or enjoyment." Dewey was a much admired and thoroughly admirable man. His influence has been beneficent in many fields. But he is not everywhere received as a thoroughly original thinker. The complaint is heard that he is turgid and prolix and that his ideas do not have that special unlocking, liberating quality that marks a first-rate philosophy.

The pragmatists are more than halfway to the analysts. Peirce regarded metaphysics a subject "much more curious than useful." One ought to know about it, he said, as one ought to know about a sunken reef — so as to steer clear of it. In the forefront of those who tried to steer philosophy away from the reefs of idealism is the famous English philosopher G. E.

Moore. Moore was a realist, an apostle of common sense and a founder of what is called the analytic movement. Among the major thinkers who were his pupils are Bertrand Russell, Lord Keynes and Ludwig Wittgenstein; all were profoundly influenced by him. The heart of Moore's philosophy is the belief that the world is made up of an infinite number of independent entities, all equally "real." Where others had established hierarchies of existence, distinctions between primary and secondary qualities, contrasts of appearance and reality, Moore saw a world in which "*everything* is real that common sense, uninfluenced by philosophy or theology, supposes real." Thus there is a place for $\sqrt{2}$ and stones; for the redness of berries and the concept of good; and no one need doubt that there are trees standing in the quad even when no one is looking. Keynes was enthralled by the "beauty of the literalness of Moore's mind, the pure and passionate intensity of his vision, *un*fanciful and *un*dressed up." A simple, childlike, repetitive quality characterizes his style — his prose reminds one of Socrates, Gertrude Stein and even Michael Arlen — and an extraordinary lucidity and austerity. When he makes a point about the meaning of the universe, neither a philosopher nor a greengrocer can fail to understand; this, if nothing else, distinguishes him from other great thinkers. It also helps to fix his place in the analytic movement, since he emphasizes the importance of intelligibility, of fixed points of reference for language, of basing knowledge on something more than internal fumes.

The largest and most imposing figure of modern philosophy is Bertrand Russell. Merely to list the subjects he has thought about, written on, contributed to, would be an immense task. He began as a follower of Hegel, abandoned his doctrines to embrace Moore's realism and then gradually relinquished Moore's views as he became more and more interested in analysis. He approached philosophy as a mathematician and logician where Moore (having had a linguistic and classical training) came to it "more like a precise philologist with an

extraordinary ear for ordinary language." White gives a succinct summary of Russell's attitude: "Instead of thinking of
mathematics, physics and common sense as axiomatically un-
touchable, instead of conceiving philosophy as essentially
spectatorial or passive *vis à vis* these more solid parts of
knowledge, Russell insists that the philosopher should enter
science and participate in the reconstruction of its foundations."

His most signal success in this clean-up job was in demonstrating how mathematics can be derived from logic. The task
involved, among others, extensions and reformations of logic
itself (George Boole, Gottlob Frege and Giuseppe Peano were
among the pioneers in this enterprise) and an assault upon
certain logical puzzles and paradoxes that had plagued philosophers since antiquity. In the course of his labors Russell
made up a few delightful and distressing puzzles of his own,
which served to emphasize how badly a reconstruction of
logical foundations was necessary. While Russell has little use
for the doctrines of such system builders as Bergson or Santayana or Croce, and while he has engaged in a "long polemic"
against pragmatism, his own conception of the role and
method of philosophy is in some ways linked with both these
tendencies. He has insisted that philosophy must meddle in
science and other activities as a critic, and must also be an independent discoverer of truth. In attempting to reform mathematics and physics, he re-evaluated and redefined the meanings
of terms "in a way that is surprisingly pragmatic." A great
part of philosophy, says Russell, "can be reduced to something
that may be called syntax." Rudolf Carnap, a prominent logical positivist, has urged the theory that all philosophical problems are merely syntactical — whence it follows that when
language is properly used, the problems are either solved or
shown to be insoluble. Russell regards this as overstatement.
Nevertheless he stresses the "very great utility" of the syntactical method. An example of its utility involves what is called
the theory of description. Phrases, says Russell, "in which a

person or thing is designated not by name but by some property which is supposed or known to be peculiar to him or it" have given a lot of trouble in philosophy. "Suppose I say 'The golden mountain does not exist,' and suppose you ask 'What is it that does not exist?' It would seem that, if I say 'It is the golden mountain,' I am attributing some sort of existence to it. Obviously I am not making the same statement as if I said 'The round square does not exist.' This seemed to imply that the golden mountain is one thing and the round square is another, although neither exists. The theory of description was designed to meet this and other difficulties." By means of the theory, Russell was able to convert descriptive phrases into a form that cut through the existence puzzle. Thus, "The golden mountain does not exist" became "There is no entity c such that 'x is golden and mountainous' is true when x is c, but not otherwise." It is important to point out that such puzzles are in fact more than puzzles — as is so often the case in mathematics and logic — and that their solution, involving as it does the elimination of errors in the use of language, has a profound bearing upon correct reasoning in all branches of thought.

Logical analysis, having sharpened its teeth on these puzzles, turned to chew upon concepts and problems of meaning in relativity theory and quantum mechanics, in psychology and in other sciences. There is, to be sure, less than universal approval of what has been accomplished in these forays into areas outside mathematics and logic, but the essential point is that the followers of logical analysis refuse to be awed by such problems, refuse to believe that there is some "higher" way of knowing. Russell, it should be added, has throughout his enormously productive life concerned himself as fully with the problems of human society — ethics, politics, education and so on — as with the problems of technical philosophy. But he has emphasized the opinion that philosophy is after truth, that it is not concerned with promoting, pursuant to certain preconceptions, "good behavior" or happiness, and that there are many questions of profound human importance that are either unanswerable or lie outside the province of philosophy.

Rudolf Carnap and Ludwig Wittgenstein are the subject of the two concluding chapters of White's survey. Carnap holds that "philosophy is nothing but the logic *of* science"; and only mathematical and empirical statements are meaningful. A statement about the outside world has to be reducible ultimately to "observable predicates," and has to be "verifiable" by some accepted experimental procedure; a mathematical statement, on the other hand, is, if true at all, a priori true — which is to say, true, independent of experience. On the view that mathematics is reducible to logic, mathematical statements are what Wittgenstein called tautologous: they assert nothing more than what is asserted in a statement of the type "All spinsters are unmarried." It must not be supposed that the insistence of the logical positivists on precise categories of inquiry, verifiability and the like cuts them off from questions of broad human concern. They are not, for example, indifferent to the study of ethics, but their approach to it is characteristic. They do not accept as part of their business norms of action or moral judgments; they are quite willing, however, to accept empirical investigations of the causes and effects of different kinds of human behavior.

Wittgenstein, a strange, lonely, moving figure, represents the extreme of the logical positivist position. An almost fanatic apostle of clarity of meaning, he himself uttered dark and aphoristic sayings, as moving, as compelling and often as difficult to fathom as the sayings of Heraclitus. He regarded *all* philosophical statements as meaningless, all metaphysics as a disease. He tried to explain the origins of this disease and how to heal it. "The meaning is the use" (which recalls Peirce) was his most famous slogan. Others of his well-known aphorisms are "The riddle does not exist. If a question can be put at

85

all, then it *can* also be answered"; "whereof one cannot speak, thereof one must be silent." Wittgenstein succeeded to Moore's chair in philosophy at Cambridge. He was much influenced by Russell. Today his own influence is at its height, chiefly as the result of the appearance of his posthumous work *Philosophical Investigations*. Since Wittgenstein regarded philosophy as nonsense, one may ask how he justified his own. The answer is, he didn't. At the end of his celebrated book, the *Tractatus*, he speaks of his propositions as "senseless" but serving the purpose of a ladder. The reader, having climbed it and gained a better view, is enjoined to throw the ladder away. Carnap and other logical positivists have felt that this is going too far. They say that Wittgenstein makes sense — an understandable position, since they are his followers.

SCHRÖDINGER—
MIND AND MATTER

H IGH up in the vault of the head, site of the
roof brain, the world comes to light. Sensa-
tions, perceptions, memories weave their images. In that tiny
tenement all experience comes to focus. Is this moving picture
a projection of a real world existing outside, or is the picture
itself the whole stuff of the world? The question has for centuries engrossed philosophers, scientists, and plain men. Those
who are called idealists have asserted that the picture is constructed by mind, and that mind is the reality and matter a
mere fiction; those who are called materialists have maintained that matter alone exists and that mind is merely one of
its properties. The meaning of the words "mind" and "matter" has itself of course been a subject of vast debate, but on

87

one point at least there has usually been agreement: Between mind and matter lies an impassable gulf.

The debate continues, and here in Schrödinger's latest volume,* the Tarner Lectures given at Cambridge University in 1956, we have yet another analysis of various aspects of the classical problem. It is not a fresh venture for Schrödinger; he trod the same ground in the epilogue to *What Is Life?*, in *Science and Humanism*, in *Nature and the Greeks*. For this restless, exquisitely sensitive thinker the mind-matter question has long had a powerful attraction. He has made his mark in physics, but his eye scans a wider horizon. He is absorbed in matters that science in its strict definition cannot answer alone. In his search for meaning — of life and the world — he has not been bound by the traditional methods of science. He has not renounced them, but he has supplemented them. Bertrand Russell wrote in a famous essay that men who have sought to conceive the world as a whole by means of thought have felt the need of both science and mysticism. Schrödinger has felt this need, and has tried to harmonize both impulses. His book is a gem with many facets; one loses oneself easily peering into its depths.

He begins his lectures by considering the physical basis of consciousness. For this purpose it is convenient to adopt the "hypothesis of the real world." Somewhere behind the eyes resides a thinking, sensing, conscious self. This is "inside." We distinguish also an "outside": an external world that consists, as we imagine, of all that has been and is — shapes, sounds, motion, light, matter. This world, we say, exists; yet existence is obviously not enough to make the world manifest. It must communicate itself to us, make itself known. We suppose this is achieved by a complex of signals, which the self interprets. How this is done is essentially a mystery, but we conjecture that certain processes taking place in a certain highly special-

* Erwin Schrödinger, *Mind and Matter*, Cambridge, 1958.

Erwin Schrödinger.
(P.I.P. photo by Ullstein Verlag)

ized piece of matter known as the brain produce certain peculiar occurrences called consciousness. All this may be nonsense, but it is not barren nonsense.

According to our model, consciousness is associated with material processes that take place in nerve cells and brains. It is well established that these biological features, which characterize many different kinds of organism, serve a unique and valuable purpose. The individual possessing such a mechanism can adapt itself, can choose, can respond to changes in surroundings by changes in its behavior. There are, to be sure, organisms such as plants that can adapt themselves in an entirely different fashion, but a nervous system crowned with a brain is not only the most elaborate and ingenious of all adaptive mechanisms, but also confers upon its owner an enormous advantage.

The brain, however, is an accident. It is a special turn of evolution that might never have occurred. Natural selection, as Darwin said, does not necessarily include progressive development, but takes advantage only of variations "beneficial to each creature under its complex relations of life." Are we to assume then that in the absence of that special event consciousness itself would not have arisen, and that the outside world would have remained "a play before empty benches"? Schrödinger finds this a repugnant idea, leading to a "bankruptcy" of the world picture. But he is not disposed to embrace the views of Spinoza and others who held that all natural objects, animate or inanimate, possess a soul. The flashing-up of the world into the light of consciousness would thus not be a privilege reserved for the higher animals, for everything would, in a sense, be aware of itself as part of a universal mind.

To extend the domain of consciousness is to court fantasies, but one can follow another path to firmer ground. Not every nervous process, nor even every cerebral process, as Schrödinger points out, is accompanied by consciousness. Descartes in a famous passage compared the body to a wound-up clock whose wheels and weights simply tell the hours by following the laws of nature. And "so in like manner [he wrote] I regard the human body as a machine so built and put together of bone, nerve, muscle, vein, blood and skin, that still, although it had no mind, it would not fail to move in all the same ways as at present, since it does not move by the direction of its will, nor consequently by means of the mind, but only by the arrangement of its organs." His doctrine has been modified but not abandoned. Mindless motor acts, it is said, constitute a large part of the activities of men and animals. We do not think to breathe, to run, to hold a glass, to swallow. The heart beats, fortunately, without its owner's attentions. Reflex actions are recognized in the vertebrate ganglia, and in that part of the nervous system under their control, which regulate and time reactions inside the system. Many reflexive processes pass through the brain that do not fall into consciousness or have very nearly ceased to do so. Schrödinger carries the point further. "Any succession of events," he maintains, in which we take part with sensations, perceptions and possibly with actions "gradually drops out of the domain of consciousness when the same string of events repeats itself in the same way very often." The boy recites the poem he has carefully memorized, and for all he is aware of its content it could be T. S. Eliot or Ogden Nash. A well-rehearsed piano sonata can be played "in one's sleep." We drive to work along a familiar route unseeing and lost in thought. There is a story of a noted mathematician whose wife found him lying in bed, the lights off, shortly after evening guests had gathered in his home. What had happened? He had gone to his bedroom to put on a fresh collar. "But the mere action of taking off the old collar had released in the man, deeply entrenched in thought, the string of performances that habitually followed in its wake."

Habit crowds out consciousness. But when something new confronts us, calling for a new response — a detour, say, on

91

the way to work, when we learn an untried action, listen, observe, explore, feel pain or joy, the signal that prods us intrudes into consciousness. Even breathing may become a subject for thought and deliberate action. Consciousness thus appears in the role of a tutor. It "supervises our education," introduces us to new facts and fresh methods, but leaves us to carry on our tried routines alone. The higher vertebrates, having the mechanism that produces consciousness, learn most easily. Their nervous system is the place, in Schrödinger's words, "where our species is still engaged in phylogenetic transformation; metaphorically speaking it is the 'vegetation top' of our stem. Consciousness is associated with the *learning* of the living substance; its *knowing how* is unconscious." Perhaps, then, consciousness is not an "exclusive property of nervous processes," but is associated "with all organic processes inasmuch as they are new." I am not sure I know what this means, but Spinoza might have welcomed it.

Now comes a diversion typical of Schrödinger. Even without the last generalization, he sees in his theory of consciousness a clue toward a scientific theory of ethics. The background of every ethical code is self-denial; an imperative "thou-shalt-not" opposes our primitive will. Commandments and principles of conduct involve a suppression of "primitive appetites," and our conscious life is necessarily a continued fight against our primitive ego. The organism is ever becoming. Every day of man's life is a small bit of the evolution of the species, as he adapts and transforms himself; and the whole evolution is the epic of a myriad of tiny transformations — "minute chisel blows," as Schrödinger calls them. We ourselves are both chisel and statue, conquerors and conquered: "it is a true continued self-conquering." Are we asked to believe that this evolutionary process — so ponderous and slow, so imperceptible not only to the individual but to entire historical epochs, so much the creature of chance mutations tested by selection — actually penetrates into consciousness? Schrödinger's answer

92

is an affirmation of faith. It is by learning, by adapting, by solving the problems created by change that we evolve, and in these activities consciousness plays a crucial role. Consciousness is a "phenomenon in the zone of evolution. This world lights up to itself only where or only inasmuch as it develops, procreates new forms." It follows that consciousness and self-discord are inseparably linked. Not the man at peace with himself, dozing like a little rabbit in the sun, but the creature of self-strife, is the hope of society. The wisest and best men of all times confirm this paradox. They suffered to achieve; for them the world was lit in a brilliant light of awareness, and only in this light were they able "to form and transform that work of art which we call humanity."

Lest the reader accuse him of preaching morals instead of expounding science, Schrödinger proffers reassurance. "Do not take it," he says, "that I wish to propose the idea of our species developing towards a higher goal as an effective motive to propagate the moral code. This it cannot be, since it is an unselfish goal, a disinterested motive and thus, to be accepted, already presupposes virtuousness." No, the "shall" of Kant's imperative remains unexplained, and the ethical law — "Be unselfish!" — is simply a fact, agreed upon even by those who do not very often keep it. Schrödinger regards its puzzling existence as an indication "of our being in the beginning of a biological transformation from an egoistic to an altruistic general attitude, of man being about to become an *animal social*." Egoism is a virtue for the solitary animal. It must be primarily self-interested to survive; in evolutionary terms, self-interest is favored by selection. But as communities arise, self-interest must be curtailed; something must be yielded to the commonweal. In certain very ancient societies — of bees, ants, termites — egoism has been entirely relinquished; the state is all that counts, and a kind of ferocious national egoism possesses the citizen. ("A worker bee that goes to the wrong hive is murdered without hesitation.") In the history of man one can discern similar developments. Whether or not we began as Hobbesian brutes, we are still "pretty vigorous egoists"; at the same time many of us are intense national egoists. Yet it has become apparent to many thoughtful persons that nationalism too is a vice. The individual's longing for peace competes with his patriotism; and perhaps the more primitive (and more rational) form of self-interest will save man from destruction. Bluntly: "If we were bees, ants, or Lacedaemonian warriors, to whom personal fear does not exist and cowardice is the most shameful thing in the world, warring would go on forever. But luckily we are only men — and cowards." Unfortunately things are not really that simple.

It may occur to you that these views of Schrödinger involve at least a tacit acceptance of Lamarckism. For if behavior plays a continuing part in evolution, must there not be an inheritance of acquired characteristics? Schrödinger was long troubled, he tells us, by this question, but he is now satisfied that he has found a way out. It consists of the operation of a kind of "feigned Lamarckism," a mechanism suggested by Julian Huxley's treatise on evolution. Lamarck's notion that the improvements or adaptations that an organ acquires through use are transmitted to the offspring is wrong. Chance variations favored by selection are accumulated, or at least accentuated; and this process, easy to describe but of inordinate complexity in its microscopic workings, can be made to account for the gross transformations of species. Yet a striking simulation of Lamarckism occurs, according to Huxley, "when the initial variations that inaugurate the process are not true mutations, not yet of the inheritable type." A specialized skill may lead an animal to change its environment, to migrate to surroundings where this skill is advantageous; thereupon the environment will not only favor those individuals most proficient in the skill but also preferentially select those in whom chance mutations have accentuated the favorable character.

Take this example. The ability to fly enables birds to build their nests high up in the trees or on inaccessible cliffs where the young especially are in less danger of being attacked by predators. Thus the ability to fly *and the use made of it* confer a selective advantage. But now the environment itself takes a hand in smiling upon the better fliers among the young. The behavior of the parents has therefore reinforced and speeded up the process of selective improvement, because when the "right" mutations show up, the individuals possessing them are so situated as to make the best use of them. We must, of course, guard against an animistic interpretation of this process, for it is tempting to lapse into metaphor and say that the species "has found out in which direction its chance in life lies and pursues this path." The fact that certain behavior can be shown to enhance the selective value of certain mutations, to pave the way for the realization of their advantages in a given environment, must not be taken as evidence that this behavior is deliberately adopted for this purpose. To avoid one's enemies if they are stronger, or to seek them out if they are weaker, is doing what comes naturally, without a grand design other than survival in view. But it may happen that just this behavior pattern, which is transmitted to offspring by example, will contribute to the selective process described above. The causal mechanism proposed by Lamarck is enormously suggestive; while it is all wrong, it is merely upside down. Behavior does not change physical organs in an inheritable way; rather physical changes resulting from mutations change behavior. This change, transmitted by example or teaching, acts as a significant evolutionary factor "because it throws the door open to receive future inheritable mutations with a prepared readiness to make the best use of them and thus to subject them to intense selection."

On this hypothesis Schrödinger is able to relate the operations of mind and consciousness to biological development. If behavior has biological consequences among plants and lower animals, where trial and error are apt to govern it, how much more important must it be in man, who can make choices — sensible choices. Political and social events are not thrust upon us; en masse, at least, we cause and control them (though we may often act unwisely). Our biological destiny is no different. Nature has not decided it for us in advance. It is possible, to be sure, that man, like the crocodile or the insects, has reached the end of the evolutionary line. Moreover, there is reason to believe we have blocked the Darwinian mechanism in both good and bad directions. We protect the weak, cure the sick, feed the hungry and practice birth control; we also slaughter each other in ever-increasing numbers, and permit millions to die of starvation and disease. Humanity and inhumanity alike interfere with natural selection. Nevertheless there remains to us a very broad opportunity to improve our biological future by cultivating the evolution of the intellect. We can hope thereby not only to enlarge our understanding of nature and our control over its forces, but to achieve mastery of ourselves and our destructive impulses as well.

Schrödinger warns of the serious danger of "a general degeneration of our organ of intelligence [because of] the increasing mechanization and 'stupidization' of most manufacturing processes." There is a constant search for talent and a cry for genius, yet the conditions of industrial society promote the rise of the cheerful robot. Schrödinger notes the controversy over the "welfare state," which is often accused of stifling incentive by leveling chances and providing everyone with economic security. But care for our present welfare need not undermine our evolutionary future. Next to want, says Schrödinger, boredom has become "the worst scourge" in our lives. Ingenious machines steadily encroach upon arts and skills; entertainment is canned and packaged; the popular image of bliss is to rot in feckless leisure, i.e., to "retire," at the earliest possible age. The machine, says Schrödinger, "must take over the toil for which man is too good, not man the work for which the machine is too expensive. . . . This will not tend to make production cheaper but those who are engaged in it happier." We place high value upon competition, but the competition of commerce and manufacture is as uninteresting as it is biologically worthless. "Our aim should be to reinstate in its place the interesting and intelligent competition of single human beings."

I shall discuss only one other of Schrödinger's themes. Two general principles, he contends, form the basis of the scientific method: The first is that nature is comprehensible; the second, that it is possible to "objectify" the world. We are concerned with the second, which the reader will recognize as the "hypothesis of the real world" mentioned earlier.

To make headway in understanding the infinitely intricate problem of nature, we pretend to be able to step out of the world, to become observers. It is clear that this is an artifice, that we are ourselves part of what we are observing, but the device appears to be essential if we are to make sense, let alone science. Yet its illogicality returns to grin at us. With the sentient self removed, we make a tidy model of a system of matter and motion. But this model has a grave flaw: it is "colourless, cold, mute," in short, it is not a model of the world at all. It is like a diagram of a flower. Having eliminated ourselves in order to get closer to the world, we have managed to lose it. Hurriedly we step back into the picture. The result is chaos. Not only is it a picture of ourselves making a picture of ourselves making a picture, but it is full of figments and fictions that came out of our heads. It is a personal world, many features of which cannot be measured and are therefore of no interest to science. My dislike of poached eggs is very real to me: Is it or is it not a proper part of the world picture, and if it is, of whose world picture?

Another difficulty. In the hypothesis of the real world, how does mind act upon matter? To make the picture, the self has been removed. But the self is mind; then how can it work upon what it is divorced from? Is this a mere quibble? Only if science and philosophy are themselves a quibble. "Physical science," Sir Charles Sherrington wrote, "faces us with the impasse that mind *per se* cannot play the piano — mind *per se* cannot move a finger of a hand." What then does what is done? Is the picture of the world of matter made by matter itself? Is there no direction, no design, no discrimination? It can certainly be argued that there is no such thing as mind, that it is a meaningless concept. This leads to one set of antinomies. But if we concede to the concept any meaning whatever, we are enmeshed in other antinomies. Mind sits alone in its high perch in a world of shadows. It is a stranger in its own world. In his great book *Man on His Nature* Sherrington epitomizes the dilemma: "Mind, for anything perception can compass, goes therefore in our spatial world more ghostly than a ghost. Invisible, intangible, it is a thing not even of outline; it is not a 'thing.' It remains without sensual confirmation and remains without it forever."

Modern science is in a trap; physics is hopelessly ensnared. Its bizarre model of the world is admittedly a model of nothing; its parts cannot be pictured; it is held together by contradictions. Time, space, matter, motion — even causality itself — have been thrown on the rubbish heap. Limits have been set upon knowledge; the physicists' Eden abounds in forbidden fruits.

All this, in Schrödinger's view, is the result of excluding ourselves from the picture. Whatever science has gained in the past from this "exclusion principle," we are now obliged to assess its consequences. It is too early to say whether science must be made anew, just as it is too early to swallow the fashionable dogma that there are certain properties of an object that cannot be accurately known. (If such boundaries to knowledge exist, we have to abandon at least partially the cardinal principle of the understandability of nature; but we are not yet, Schrödinger feels, forced to this desperate step, because physics may find more plausible models.) It is not too early, however, to criticize the contention, based on recent discoveries in physics, that refined methods of observation have carried us so close to the mysterious boundary between subject and object that the boundary has begun to vanish. Schrödinger's demurrer to this contention is characteristically subtle. Many thinkers have, of course, made the point that the observer colors the observation, that what we perceive is not the "thing in itself," to use Kant's term. But modern physics goes beyond this in asserting that it is not only our impressions of the outside world that depend on our "sensorium," but the outside world itself that depends upon it and is changed by it. Yet how, asks Schrödinger, can this be? The clumsy finger may tilt the scale, the eyelash may blur the image under the microscope, but how can the subject's mind, the thing that merely senses and thinks, disrupt the physical world? The mind is not matter, nor does it belong to the world's energy. This is the core of the real-world hypothesis, the root of objectivation.

There is, however, another approach to the dilemma. We need not, after all, accept in the first instance the "time-hallowed distinction" between subject and object. It has its practical uses, and in science it serves as an invaluable make-believe; but in philosophy, as Schrödinger believes, it should be abandoned. Scientists do not get into trouble when they act upon the distinction in their own field, but rather when they decide to draw portentous epistemological consequences from their unsolved problems. This does not mean that the scientist should stick to his last; Schrödinger is scarcely in a position to advocate such self-restraint. But it does mean the physicist must clear his thoughts, must recognize that pragmatic assumptions in one sphere do not necessarily yield valid conclusions in another. Let Schrödinger summarize: "It is the same elements that go to compose my mind and the world. This situation is the same for every mind and its world, in spite of the unfathomable abundance of 'cross-references' between them. The world is given to me only once, not one existing and one perceived. Subject and object are only one. The barrier between them cannot be said to have broken down as a result of recent experience in the physical sciences, for this barrier does not exist."

I do not think it can be said that Schrödinger has made a solid contribution to the science of mind; and it would be useless to pretend that a fine linguistic scalpel, of the sort the analytic philosophers have devised and are so mercilessly skillful in wielding, is needed to slice many of his arguments into bits. He is old-fashioned in some of his views; his concepts are apt to leak and flow into one another; he is sometimes, I regret to say, fervid when it would be better to be clear. And yet he has given us something that stands nobly by itself, that can survive the onslaughts of both scientists and philosophers. What he has written is imbued with values that are as compellingly real as they are hard to define. There is not a trace of superficial cleverness about him. To a fine intelligence he joins a passionate heart. He is not ashamed to be human, though he is never sentimental. That the dilemma of objectivation is his primary concern is a reflection of his whole style and outlook: He puts himself into his ideas, he shares his doubts with us, he pierces us with his reflections. For Schrödinger the importance of the questions of science lies in the fact that they are questions about the nature and meaning of life. Because these questions cannot be answered by science alone, he has turned elsewhere for light. The book of Job tells us that the ways of God are inscrutable, that there are questions we must not ask, things too impious to search out, "too wonderful" for man to know. For Schrödinger, if the question can be put, the answer must be sought. One can read his little essay in a few hours; one will not forget it in a lifetime.

100

THE SCIENTIFIC ATTITUDE

C. H. WADDINGTON, professor of animal genetics at the University of Edinburgh, is of that small group of scientists — may their tribe increase — who understand the relation between science and society and have a sense of responsibility for social problems that transcends the normally limited outlook of the specialist. They are, in other words, sensible and courageous men as well as scientists. The conjunction of these qualities is rarer than is perhaps commonly supposed; and on this side of the water, especially, as the pressures of conformity mount and the penalties for social and political heresy become severer, the number of scientists at once aware of their time and prepared to defend sane ideas and rational behavior outside the domain of their disciplines grows dangerously small.

Waddington's little book, *The Scientific Attitude*, written for the admirable Pelican series, struck me on its appearance in 1941 as the work of an uncommonly perceptive and stimulating mind. Revised and somewhat enlarged, the book has now been reissued* and on rereading it I find my original impres-

* Pelican Books, London.

sions reinforced. Because the issues it treats have reached an even more critical stage than in 1941, I should like to discuss a few of its points at some length.

Waddington began his career as a geologist interested in the historical aspects of evolution; later he turned to genetics, where he has attained first rank. He is a fellow of the Royal Society, has published many technical papers and several books, including *Science and Ethics*, and keeps watch over social and political matters no nice biologist ought to worry about. To this essay on the common sense of science, Waddington brings a happy mixture of skepticism, tolerance, scientific insight, friendliness and a practicality gained, at least in part, from his first-hand experience in operational research with the British Air Force during the war. Here, it is plain to see, is a scientist who respects men as well as ideas and has taken thought to both; he is what is certainly exceptional — a scientist with a scientific attitude.

What is the ideal scientific attitude? I would find it not easy to describe, but fortunately Waddington's book makes a good crib. The scientific attitude is, of course, less the product of innate qualities than of education and social influence. Its attributes depend on present circumstances and are neither ubiquitous nor eternal; the scientific attitude suitable to the age of Roger Bacon or to the Balinese is not the scientific attitude appropriate to twentieth-century Western man. It makes demands — and affords rewards — in accord with our cultural heritage. It is unsentimental, unsuperstitious and undogmatic. It pledges no irrational loyalties, is free of the bias of property owners and wears neither school tie nor political badge. It is as distant from the crowd as from the cloister. It is not — the point will come up again — the censor of any faith so long as that faith is not the censor of the normal appetites and reasonable hopes of the mass of men. It is curious, sympathetic, open, tentative and hospitable to change.

It is the friend of freedom and the enemy of its enemies. It is neither cynical nor despairing, nor is it shattered by the unexpected. It is based on the awareness that men, though at the apex of the Linnaean chart, are animals and not seraphs; it asserts that they have animal needs, which must be met by using the resources of the physical world, and that these needs can be better determined and satisfied by rational scientific methods than by other methods (e.g., that a belief in the efficacy of crossing oneself, or in the virtues of war or nationalism or of a particular economic system, is not among the better ways of satisfying these needs). It is frankly pragmatic, seeking, in Macaulay's words, "not to make men perfect but to make imperfect human beings comfortable." Its yardstick for measuring the "good" in the absence of any other adequate standards is the "general course of evolution."

An irreproachable set of attributes and principles. But can they be applied effectively to politics as well as physics, to social action as well as research, to practice as well as thought? That question presses upon us for reasons familiar enough: because men's ability to order their affairs sensibly has been outstripped by their ability to make machines; because machines, when misused, not only debase and brutalize life but now in truth threaten human survival; because the scientist of our period is often no more than a technician unwilling (and unable) to extend his methods beyond a narrow area, to explain the implications of his discoveries.

Take the last charge. There is, to be sure, a popular notion that "science is and must be ethically neutral," that the scientist should stick to his last. Convention frowns upon the crystallographer concerned with the social function of science, the astronomer who debates foreign policy, the chemist who writes on the social implications of atomic energy, the mathematician interested in Marx, or the physicist who utters opinions on the economy or the arms budget.

Does the scientific attitude endorse this stand and require the scientist not to "meddle," as Hooke cautioned the Royal Society, with "Divinity, Metaphysicks, Moralls, Politicks, Grammar, or Logick . . ."? The question is rhetorical. It re-

103

quires no elaborate argument to prove that the scientist with the scientific attitude, the responsible scientist, must be a meddler. The tendency to regard the scientist as defiled if he mixes in social and political activities arises from the widespread failure to understand that science lies at the center of the network of human affairs. Since everything science discovers affects man's behavior and brings about changes — slight or profound, immediate or deferred — in his relation to himself, society and the physical world, it is evident that science has ethical consequences and that the notion of a "neutral," passive science is absurd. There is no inconsistency between the proposition (which is true) that the scientist at work must not permit his preconceptions and preferences to interfere with an unbiased appraisal, and the proposition (which is equally true) that the scientist has the right, and duty, to express opinions about the social meaning of his results.

The responsible scientist is ethically obligated to be unneutral, to direct attention to evils when he sees them, to press for reforms when the results of his research point the need and the proper direction; in sum, to regard his special knowledge not as a disability but as a means to special good. There are those, to be sure, who will argue that the scientist already speaks with an excess of authority and that to broaden this authority further means abandoning our society to the high priests of technocracy, to men who place efficiency above all other human values. One way to meet this false charge is to explain the kind of person the scientist is. An accurate portrayal will add up to something like this: He is neither the Wizard of Oz nor the benign doctor in the baby-food ads; he enjoys normal activities, hobbies, prejudices, political views and a love life; he is as well-fitted as other educated men to play an important part in the evolution of general culture. Waddington says:

It is time that scientists became willing to state explicitly that the scientific attitude is as full of passion, as much a function of the whole man and not merely of an intellectual part of him, as any other approach to human action. It differs from them only in what it is trying to do. Instead of trying to earn more money, or to improve the condition of the working class, or to create visual beauty, a scientist tries to find how things work. The search for causal connections cannot be made merely by refusing to grind axes. Scientific imagination and insight do not automatically result when the mind is swept clean of preconceived notions and prejudices; their attainment is a positive achievement and not a merely negative one. And because this is true, scientists can and do pass ethical judgment on human behavior; those things which are based on the scientific attitude, or encourage it, are good, those which stultify or deny it are to that extent bad.

A crucial testing ground for the broad validity of the scientific attitude is the field of politics. Even if there is something profound in politics that, according to the maxim mocked at by Swift, "men of plain honest sense cannot arrive to," science cannot avoid grasping the nettle. Science recognizes, however, that in politics it faces problems more difficult than any that arise in the comparatively orderly world of physical matter. What a trivial exercise, you will agree, to construct a rocket for interstellar travel, compared with the task of explaining the behavior, say, of Jersey City's electorate.

What is needed for an improved understanding of political (as well as economic) behavior is a systematic, impartial analysis of society's aims and beliefs. The first step is to discover what people believe; the second, to relate their beliefs to their actions; the third — and the most precarious — to supplant irrational and inconsistent beliefs, from which stem unsuitable actions, by a set of rational and consistent beliefs from which fruitful action may develop. A formidable procedure, yet necessary, if reason is to dispossess outworn standards, false values and conflicting shibboleths.

As an example of the "disparate and confusing cultural environment" in which the members of our own society are forced to live, Waddington quotes the social antinomies in the famous tabulation by Robert Lynd: Individualism is the law of Nature and the "secret of American greatness, and restrictions on individual freedom are un-American and kill initiative" — *but*: "no man should live for himself alone, for people ought to be loyal and stand together and work for common purposes." The family is "sacred" and "our basic institution" — *but*: business is the lifeblood of the nation and all other institutions must conform to its needs. "Religion and the 'finer things of life' are our ultimate values" — *but*: "A man owes it to himself and to his family to make as much money as he can." Honesty is the best policy — *but*: "Business is business, and a business man would be a fool if he didn't cover his hand." Etc.

It is true that contradictory beliefs are held in every society. But, Waddington points out, "the contradictions have never appeared so obvious, have never been so much in the forefront of consciousness and caused such mental strain as they do to-day." It is the peculiar business of science, which by some of its own works has undoubtedly aggravated the strain, to sweep out this mass of rubbish, to help create a biological and social environment in which men will not be forced to run about like laboratory animals in a maze constructed to test their mental breaking point.

Apart from the polygonal assault that the co-ordinated natural and social sciences could make upon the problems of society, the scientific attitude would destroy many of the irrational beliefs underlying social ills. Its wide adoption as a creed, the practice of its articles with the same sincerity as we practice those of strange and less reasonable creeds, would lead men, among other things, to assert their individuality, to put forward their own points of view, to distrust authoritarianism, to scorn conformity. It would train them to support their social and political views "by reasons which other people can verify," to "accept the judgment of critical experiments" as to whether they have made out their case. In daily affairs the scientific attitude, as Waddington emphasizes in one of his sharpest passages, is the "recognition that one belongs to a community, but a community which requires that one should do one's damnedest to pick holes in its beliefs. I know of no other resolution of the contradiction between freedom and order which is so successful in retaining the full values of both."

What is the relation between the scientific attitude and religion? This is more, as Waddington's essay shows, than an academic question. For although it is commonly supposed we live in an irreligious age, and it is true that religion has been thoroughly fragmented and many modern faiths so diluted that even Voltaire could swallow one of their sermons without ill effects, our immediate period has witnessed a remarkable, and in some respects alarming, revival of religion. The authority and prestige of science have been invoked to prove that God is present in the laboratory; the overthrow of strict physical determinism, having, it is said, freed us from the prison of predestination, encourages the most extraordinary theological speculations. There has also been an abundance of mischievous, politically motivated appeals to orthodoxy in religious as well as social matters, appeals invoking fear and superstition and inviting a headlong retreat from reason.

The scientific attitude recognizes that men hunger for certainty, that in a troubled world particularly they need a secure system of beliefs to which they can turn. "If men cannot live on bread alone," wrote Whitehead, "still less can they do so on disinfectants." Any attempt, it is highly probable, to make skepticism and empiricism the basis of an entire society's philosophy must fail: in Russia, for example, the Holy Fathers of the Revolution have become necessary substitutes for the Holy Fathers of the Orthodox Church.

Science is in no position easily or quickly to satisfy this hunger. It sees a constantly changing world whose most positive attributes are only expressible in the fractions of probability. It would not venture to contradict the philosopher who asserts that the ultimate nature of experience is beyond analysis. It offers no rosy assurances to the living and does not pretend it can improve the condition of the dead. For an unhappy and bewildered man the qualified, nervously circumspect propositions of science are a poor substitute for the eternal truths of religious faith.

Here I should turn aside to point out that the scientific attitude is under further handicaps. Not the least of these stems from the fact that science is itself a kind of faith and thus has to get itself accepted. The faith in science as a means of understanding and controlling nature is not automatically accompanied by a faith in science as "prototype for all human common action." Because men believe in science, i.e., electric blankets or the multiplication table, it does not follow they believe in the scientific attitude toward society. And it is precisely in this sphere that science clashes with some of the opinions of organized religion, particularly those of the more irrational faiths. Thus it comes about that its own ethic requires the scientific attitude to assume the role of a fighting and competing faith.

Yet this is a role for which, partially because of its rationalism, the scientific attitude is peculiarly unfitted. It has no holy wars at its disposal; on crucial issues its methods of inference would be judged by many religions inferior in authority to the procedures of revelation; it cannot, even apart from this challenge, point confidently to the "you must" of logic, since logic, as the mathematical philosophers inform us, is propped on stilts of arbitrary choice, planted in a bog of paradoxes. Nor can the scientific attitude without jeopardizing its integrity accept the fashionable compromise which says: "I have faith in reason for some things — those I think I understand — and faith in religion for other things — those I think I do not understand": an eclecticism as commendably complete as that practiced by the celebrated Wilson Mizener, who on his deathbed, it is reported, had a spiritual send-off administered by ministers of at least three different faiths, on the ground that this was no time to take needless risks.

The scientific attitude, we are bound to conclude, has no means of winning adherents other than by the pull of its inherent plausibility and intellectual integrity. Science must make its way by proving not only that it is the source of material progress but also that it is able "to provide mankind with a way of life which is . . . self-consistent and harmonious." The scientific attitude acknowledges the value of moral convictions, which give direction and encouragement to all human activities; of ideals, found in religion and elsewhere, which supplement the contribution of science to human betterment. It must, on the other hand, expose those aspects of religion which maintain its authority by the use of fear, superstition and intolerance, and denounce alike the alliance between reactionary religion and fascism, and the subordination of scientific freedom to the philosophical and political orthodoxy of communism.

The charge is often made that science does not impart the warmth, the intimacy and the color of life. For these values one must, Waddington would agree, turn to other activities such as art, literature, poetry, even philosophy. Yet for all its cautiousness and "heavy vernacular," science offers its own "thrill of romance." Almost invariably, fundamental scientific discoveries entail the overthrow of cherished, established beliefs. Every major theory — of curved space, of the uncertainty of causal relations, of the quantification of energy, of the particulate nature of inheritance, of the endless continuity of living substance, of the equivalence of mass and energy — warily as it may have been advanced, is more "outrageous" in its implications than the furthest flights of religious or mystical imagination. Science has adventures and fantastications for every taste.

Though the scientific attitude considers every dogma to be a source of evil, it cannot fail to notice that the certainty offered by orthodox religion is much less of an "intellectual and moral drug" than the certainties of racism, nationalism, religio-fascism and their kin. These are the wellsprings of tyranny, sustenance for seekers of power in the social, political and economic spheres. We are familiar also with the critics, for whatever purpose, of centralization, social planning and of every aspect of the welfare state, who, in their arguments, invoke the certainties most destructive of freedom as allegedly the sole means of preserving it. Hayek, for example, is typical of the school that urges men to accept social and economic institutions as they are, or at least as they were in the halcyon past, and not to keep questioning and nagging to get at the reason of things. These unsettling practices lead, says the Austrian *Gelehrte*, directly to serfdom.

Science is not easily misled by such hokum. Unquestionably, Waddington says, science is "on the side of democracy"; but only a democracy that is prepared to make social experiments and to abide by their results. In such social experiments the scientist's function is to suggest direction and method, while the members of the society judge the results and decide whether the experiment should be continued. This, in a sense, represents the ideal union of science and society.

For all its length, my review touches, and then only superficially, on no more than a small fraction of the topics considered by Waddington. I hope, notwithstanding, to have conveyed the taste of a gratifying intellectual experience.

"Consider the fact," says the Arch-vicar of the new society in Huxley's *Ape and Essence*, [that] "nobody wants to suffer, wants to be degraded, wants to be maimed or killed. [Yet] at a certain epoch, the overwhelming majority of human beings accepted beliefs and adopted courses of action that could not possibly result in anything but universal suffering, general degradation and wholesale destruction. The only plausible explanation is that they were inspired or possessed by an alien consciousness, a consciousness that willed their undoing." Consider the fact that this is the sort of diagnosis intellectuals of the Huxley-Waugh school are now peddling; you will understand the need for the scientific attitude, and the importance of Waddington's book.

PART
2

CAUSALITY AND
CHANCE IN
MODERN PHYSICS

"At certain periods in the development of human knowledge, it may be profitable and even essential for generations of scientists to act on a theory which is philosophically quite ridiculous." — C. D. BROAD

I N 1952 the physicist David Bohm published in the *Physical Review* two articles proposing a causal reinterpretation of quantum physics. Louis de Broglie had made the first approach in this direction some twenty-five years earlier, but criticism of his paper in the *Journal de Physique* led him to abandon his theory. Bohm has since revived and enlarged it. He has suggested ways to overcome its major difficulties and has invented certain ingenious argu-

113

ments and supplementary features. These results have encouraged De Broglie to take up again his original concept. Moreover, one of his younger colleagues, Jean-Pierre Vigier, has collaborated in research with Bohm and together they have achieved a fresh interpretation of the statistical significance of Erwin Schrödinger's famous Ψ function in wave mechanics. In short, a small but gifted and determined group of investigators is devoting its efforts to a re-examination of the ruling doctrine of contemporary physics. Re-examination is perhaps too polite a term. The members of this group (which includes other physicists than those mentioned) are profoundly dissatisfied with the prevailing belief; they have a doctrine of their own, which they hope will help to disenthrall the majority. In De Broglie's words, the aim is to "rescue quantum physics from the cul de sac where it is at the moment." The struggle is not without drama.

The rebel's case is brilliantly argued in Bohm's book.* He begins with a philosophical prologue on cause and chance, which I shall pass over. Though I suspect it is dear to his heart, it is the least convincing section of his brief. Not so the historical survey. Following Bohm's lucid recital — the penetrating judgments, the original analysis of the growth of ideas and their interaction — one sees how his interpretation of the history of physics has shaped his theories. It is as a historian, more than as a philosopher, that he best explains himself.

To understand the problem of causality and chance in modern physics, one must review the status of these notions in classical physics. A few landmarks must suffice. We may regard Newton's laws of motion as the supreme model of mechanistic determinism. These laws imply that, given the initial positions and velocities of the bodies in a system at a given instant of time, and the forces acting upon them, the future behavior of the system is determined and can be predicted for all time.

* David Bohm, *Causality and Chance in Modern Physics*, New York, 1957.

In brute experience this is a fairy tale. There are no isolated systems. Our information about positions, velocities and forces is never complete. But in planetary astronomy the laws work very well, and there was in Newton's day no reason to suspect that they would not work perfectly both up above and here below.

The Marquis de Laplace, it will be recalled, converted this principle of mechanics into a universal principle. Everything in the world, he said, obeys Newton's laws. They applied, one might assume, to one's disposition and to the housemaid's knee as well as to a cannon ball and the water in a river. The universe consists of nothing but bodies moving through space. Democritus and Leucippus, the Greek atomists, had had the same idea but lacked the advantage of Newton's insight; his laws, Laplace believed, made it possible to conceive of the world as a machine that is frictionless, can never break down or wear out, and, having once been started, runs on forever in predestinate ways.

In this form the mechanistic philosophy was both a deterrent and a boon to further scientific progress. It narrowed and hardened thought (much as the indeterminacy principle does today); it promoted, on the other hand, a more quantitative, unified and dynamic point of view than had before flourished in science.

But the path of the mechanists was not smooth for long. Three major advances in physics shook confidence in Laplace's dream. The first was electromagnetic field theory. As men began to understand electricity and magnetism and strove to formulate the basic laws of these phenomena, it became apparent that the Newtonian scheme needed to be supplemented. Newton's bodies occupied a definite region of space. (In some mysterious fashion, to be sure, their influence vaulted over space and acted at a distance, but the less said of this the better.) Electricity and magnetism, however, were incorporeal and could not, like butterflies, be fixed with pins. A new ap-

115

purtenance had therefore to be added to the classic model, namely, electric and magnetic fields. These were defined as continuously distributed throughout space as a whole, and at each point in space at each instant of time the components of the field were assumed to have definite values. Electric and magnetic fields are not independent: a moving charge is a current and produces a magnetic field; a magnet in motion produces an electric field. Moreover, fields act upon bodies. Fields and bodies therefore codetermine each other. To describe the traffic of the universe, a combined set of laws was necessary: Newton's equations of motion, and Clerk Maxwell's equations, which determine the interrelations of field quantities, fields, and the motions of the charged bodies in the system. Mechanism was no longer simple.

Additional abstractions made it terrifyingly intricate. For a time imagination clung to the ether as the vehicle for electromagnetism; the Michelson-Morley experiment extinguished this solace. Since the field appeared to exist quite happily in empty space, carrying energy, momentum and angular momentum, it could simulate some of the properties of moving bodies. Einstein went further; he suggested that special kinds of fields (satisfying nonlinear equations) might exist, "having modes of action in which there would be pulse-like concentrations of fields, which would stick together stably, and would act almost like small moving bodies." It is possible, he thought, that the fundamental particles of physics consist of nothing but such agglutinations.

Mechanism is a fighting faith. It did not succumb to field theory. There had been a miscalculation, certain parameters having been overlooked. But this was not fatal. If the whole of nature could not be reduced to the motions of a few kinds of bodies, it could be reduced to the motions of a few kinds of bodies plus a few kinds of fields. (Or, if one followed Einstein, to a few kinds of fields alone.) Laplace's superman would merely have to include the infinity of variables of the field in his other calculations. He would then have no difficulty predicting future states. Mechanism is also a supple faith.

The molecular theory of heat and the kinetic theory of gases posed a second challenge by introducing a qualitatively new aspect of the laws of nature. They forced recognition of the fact that what appear to be smooth, predictable events on the macroscopic level are in truth statistical averages of a vast aggregate of irregular, unpredictable events on the microscopic level. The mean pressure, for example, of a gas on the walls of a container depends almost entirely on the general properties (e.g., mean density, mean kinetic energy) of the gas molecules as a group, and is insensitive to the motions and arrangements of the individual molecules. In the small there is an enormous number of irregularities and fluctuations, but on the average these offset each other and cancel out, the result being practically determinate mean values.

It is worth taking a moment to weigh the significance of this development. If we picture the universe as a large Newton-Laplace machine, we imagine any given sequence of motions as causally and mechanically linked. The corollary of this notion is that there is a continuity of the causal chain such that any effect is directly traceable to the events that produce it, however numerous, remote or complex. New factors, however, are introduced by the kinetic theory, which enrich the conceptual structure. For instance, Bohm says, one is justified "in speaking of a *macroscopic level* possessing a set of relatively autonomous qualities and satisfying a set of relatively autonomous relations, which effectively constitute a set of *macroscopic causal laws*" — the laws of thermodynamics and macroscopic physics in general. As illustration, consider the properties of a liquid. It flows; it wets things that it touches; it tends to maintain a certain volume. Its motions satisfy a set of basic hydrodynamic equations whose variables and constants express only large-scale properties such as pressure, temperature, density, velocity and the like. Now, if we are interested

117

in describing or predicting the behavior of a mass of water, we do not treat it as a crowd of molecules, but as an independent macroscopic entity, following laws appropriate to the macroscopic level. Does this mean that the properties and behavior of the water are independent of those of its molecular constituents? Not at all. But it does mean that the large-scale entity is not only insensitive to individual molecular gyrations, but also has properties peculiar to itself, which the molecules, under certain arrangements, create but do not themselves possess. Bohm offers this example: When a liquid has a certain density, the intermolecular forces are in balance. If the density is increased or lowered, repulsive or attractive intermolecular forces manifest themselves, which, in either case, tend to bring the density back to its original value. With the forces in balance, the liquid exhibits a stability of behavior; this stability is characteristic of macroscopic entities and is not an attribute of molecules.

Another point bearing on the mechanistic philosophy is the clarification of the relationship between qualitative and quantitative changes. Take the transformation from gas to liquid to solid. In a gas the molecules are in perpetual, chaotic motion. As the temperature is lowered, clusters of molecules begin to form, but they break up almost immediately because the mean kinetic energy is high. But as one approaches a certain critical temperature, the clusters build up, on the average, faster than they disintegrate. Small droplets appear and the liquid phase gains the upper hand. Now the substance no longer fills the entire space available to it but occupies a certain characteristic and definite volume. A further drop in temperature is accompanied by an increase in density and viscosity. The atoms start to arrange themselves in regular and periodic lattices, and crystals take shape. At first these too have a short life because of the disruptive effects of random thermal motion, but again, around a certain critical temperature, the intermolecular forces achieve the most stable possible configuration, namely, the crystal lattice, and the liquid becomes a crystalline solid. In this state the substance tends to maintain a fixed shape, to transmit and to polarize light, to exhibit its own special X-ray diffraction pattern.

It is clear, then, as Bohm states, that a series of quantitative changes — in the mean kinetic energy of molecular motion — have produced a series of qualitative changes in the properties of matter. Moreover, these new properties are subject to "new kinds of causal factors . . . which 'take control' of a certain domain of phenomena, with the result that there appear new laws and even new kinds of laws, which apply in the domain in question."

The third major advance to prompt a re-evaluation of classical mechanism was the introduction of probability and statistics in connection with the analysis of Brownian motion and the formulation of the laws of thermodynamics. The extension of insights embodied in the kinetic theory of gases, and the development of the mathematics of probability, greatly enlarged the understanding of the relationship between small-scale and large-scale phenomena. By a brilliant stratagem a pyramid of knowledge was erected on a platform of ignorance. Men accepted the chance character of events on the microscopic level; but chance, as they discovered, has its rhythms and symmetries. Once it was learned how to express them in symbols and to incorporate them into a calculus, it became possible to design reasonably determinate laws for the mean behavior of large aggregates.

"It seems probable to me," Newton wrote in his *Opticks*, "that God in the beginning formed matter in solid, massy, hard, impenetrable, movable particles, so very hard as never to wear or break in pieces, no ordinary power being able to divide what God Himself made *One*, in the first creation." So long as this belief persisted, the mechanistic philosophy was enormously persuasive; indeed, it had no reason to fear a competitor. Even after confidence in it had been eroded by the nineteenth-century advances in physics, ingenious arguments were found to support the mechanistic creed. Admitting that the particles of matter are neither solid, massy, hard, nor impenetrable, that atoms are not the basic building blocks, and that every time a new particle is discovered it turns out to be a changeling, it is still reasonable to suppose that a fundamental law rules the universe. From this law it will be possible to deduce the properties of hyperons, atoms, molecules and everyday matter.

Yet suspicion grew that mechanism was coming to the end of the line. The classic causal laws leaked like a sieve. Every attempt to formulate a precise description of behavior at a given level was disturbed by apparently random events at the level below. There was no known case of a causal law that was completely free from dependence on "contingencies introduced from outside the context of the law in question." Moreover, even if such a law could be imagined, the problem would arise "of reciprocal relationships between levels and between qualitative and quantitative laws."

At the beginning of the twentieth century a new theory was advanced to escape this dilemma. "Indeterminate mechanism," as Bohm calls it, represented a momentous concession. For the first time the notion was entertained of an element of "absolute arbitrariness and lawlessness" entering into physical events. It was possible still to picture the universe as a great machine, but it was a machine more like an "idealized roulette wheel" than a perfect watch. It was the very nature of the wheel to be unpredictable in individual results, yet on the average to yield quite regular distributions. In studying natural phenomena one might expect at every level to encounter irregularities and chance fluctuations. The farther one probed, the more one might learn of causes and motions, but at bottom was an irreducible arbitrariness. This did not, of course, extend to statistical aggregates: and thus was science rescued from logical bankruptcy. To seek a fundamental, purely quantitative law that described the working of the universe remained a reasonable objective; but the law when found would necessarily be probabilistic and not deterministic.

It is interesting to observe how one dogma was advanced to replace another. Deterministic mechanism held that chance was merely the name of ignorance; that all was ruled by iron laws and iron linkages; that once these were fully known, statistics would no more be needed to explain the universe than to help a man ascertain how much money he has in his pocket. Indeterministic mechanism (as developed by the late Richard von Mises, and supported by the proponents of the usual interpretation of the quantum theory) held that all determinate laws are nothing more than "approximate and purely passive reflections of the probabilistic relationships associated with the laws of chance"; and that individual processes and events in the atomic domain are "completely lawless." The second dogma is even more unconvincing than the first. Mechanism of the first kind has no experimental basis; as a philosophic assumption, however, it is not at odds with the scientific spirit. Mechanism of the second kind also has no experimental basis, but as a philosophy it lies somewhere between a *credo quia absurdum* and a firm belief in industrious angels.

The quantum theory, says Bohm, was the first example in physics of "an essentially statistical theory." It did not assume that very small things obey the laws of classical mechanics, to which statistical considerations are applied for practical reasons. Instead, it restricted itself entirely to statistical predictions, "without even raising the question as to what might be the laws of the individual systems that entered into the statistical aggregates treated in the theory."

And now physics took the final step in breaking with the past. Heisenberg's indeterminacy principle can be interpreted as a flat denial of causality in the atomic domain. It does not

121

merely state that the causal links at this level are beyond man's power of detection; it clearly implies that the links do not exist. This was Heisenberg's own inference from his principle, an inference to which the majority of physicists subscribed. As to indeterminacy, three points are worth mentioning. The first is that Heisenberg's hand was strengthened by a theorem of the late John von Neumann, published in 1932. This theorem appeared to furnish a mathematical proof that it is impossible to conceive of a distribution of motions of "hidden" parameters that would account for the behavior of an individual system at the atomic level; which is to say that, even if information about a system could be got without disturbing it by measurement, precise predictions about its future could not be made. The second point is that the usual interpretation of the quantum theory forsook the concept of continuity of motion, as well as causality. (This, by the way, offered a roundabout solution of Zeno's puzzles.) The third point is that Niels Bohr introduced his principle of complementarity to preserve the logic of physics from ruin. The indeterminacy principle had ostensibly ruled out precisely definable conceptual models; Bohr proposed as a substitute the use of complementary pairs of imprecisely defined concepts: position and momentum, wave and particle, and so on.

That indeterminacy is at the heart of things is a view for which a number of distinguished physicists, including Albert Einstein and Max Planck, felt no sympathy. De Broglie and Bohm share this skepticism and are in fact convinced that there are serious flaws in the reasoning underlying the prevailing conclusions. Ernst Cassirer, in his excellent book *Determinism and Indeterminism in Modern Physics*, rejected the idea that indeterminacy dooms causality; Bohm goes a good deal farther.

It may be, he says, that while the quantum theory, and with it the indeterminacy principle, are valid as high-degree approximations in a certain domain, "they both cease to have relevance in new domains below that in which the current theory is applicable." Accordingly, one must try to construct a new model incorporating a "sub-quantum mechanical level." If, with the help of such a model, all the phenomena can be explained that quantum mechanics explains; and, in addition, phenomena can be explained that quantum mechanics is unable to explain, a momentous break-through will have been achieved. It is to this hypothetical structure that he has turned his mind.

His main argument runs something like this. Assume a lower basement of the physical world, not yet discovered; assume, further, that hitherto "hidden variables" are at work in this basement. We may then conjecture that the statistical character of the quantum theory arises from random fluctuations of new kinds of entities existing at the lower level. Indeterminacy will in that case be a necessary attribute of phenomena at the quantum-mechanical level, simply because the motion of the entities that can be defined at that level alone are determined by factors (hidden variables) that cannot be so defined.

Does the indeterminacy principle, as such, rule out the possibility of a sub-quantum-mechanical level as merely an "empty metaphysical speculation"? No, says Bohm; for the principle has "nothing whatever to say about the precision that might be obtained in measurements that utilize physical processes taking place at a lower level." Is there any reason to believe in the existence of such processes? Bohm suggests they may be found in the domain of very high energies and of very short distances. This is the very domain in which quantum mechanics has exhibited serious shortcomings as a tool of investigation, thus leading to a "crisis" in microscopic physics. What of Von Neumann's theorem, which can be interpreted to mean that a lower level is theoretically impossible? Here the refutation is directed to what Bohm regards as a defective assumption; namely, that in specifying the state of a system, one cannot dispense with the aid of "observables" that satisfy

123

certain rules of the quantum theory. From this Von Neumann went on to prove that no possible set of hidden variables would help describe the system more precisely than the current formulation of the quantum theory. But the assumption, according to Bohm, begs the question; for if a sub-quantum-mechanical level exists, it is not unreasonable to suppose that the entire scheme of "indispensable" observables is inappropriate to this level and must be replaced by something very different. And with the observables go the infirmities peculiar to the quantum theory.

Imagine the following model. Connected with each "fundamental" particle of physics is "a body existing in a small region of space." The body can be conceived as a mathematical point. Inseparably associated with the body is a wave, assumed to be an oscillation in a new kind of field. The field is represented by Schrödinger's Ψ function, which now suddenly comes to life and, instead of being a mere symbol for the calculation of probabilities, represents something as real as the electromagnetic and gravitational fields. Between the Ψ field and the body there is interaction: the field exerts a "new kind" of quantum-mechanical force on the body, which is noticeable at the atomic level but not above it; and the body exerts a smaller but definite influence on the field. The nature of the field force is undefined, except that its tendency is to pull the body into regions where the value of the field is largest. There is, however, a resistance to this tendency, consisting of random motions of the body (analogous to Brownian motions). A possible source of these motions is random fluctuations in the field itself. (Electromagnetic fields, for example, undergo similar irregular oscillations.) One may suppose that the fluctuations are associated with properties of the field at a *sub-quantum-mechanical level*; or arise from interaction with entities existing at this lower level. On this hypothesis, there are two forces at work; the fluctuations that cause the body to wander at random over the whole space accessible to it; and

124

the quantum-force that pulls the body to where the field is most intense. The result is a "mean distribution in a statistical ensemble of bodies," corresponding to Max Born's probability distribution. But notice that instead of accepting the distribution, as does Born, as an absolute and inexplicable property of matter, one can interpret it as the effect of sub-quantum-mechanical-level random motions.

Bohm claims that the model yields results consistent with all the essential results of quantum theory. Take the business of wave-particle duality. A famous example from quantum theory shows that when electrons pass through a pair of slits and fall upon a screen, an interference pattern is formed, as well as a series of discrete dots. Moreover, the closing of one of the slits appears mysteriously to influence even the particles that pass through the other. The cause of this anomaly is a forbidden subject in the usual interpretation. But the new model yields an explanation. The interference pattern is produced by the waves associated with the electron. The small body connected with the electron undergoes random motion and follows an irregular path. But a large number of these bodies passing through the slit system produce a statistical pattern of dots at the screen, whose density is proportional to the field intensity. The "quantum-force," in other words, accounts for the concentration, and the random motions (of lower-level origin) for the irregularity, of the array of particle images. And the enigmatic closing effect is assumed to mean that closing one slit influences the "quantum-force" acting on the particle as it moves between the slit system and the screen.

Bohm considers certain of the criticisms of this model related to problems of electron spin, the theory of relativity, and the basic implausibility of the postulated interaction of wave and particle. He gives an account of the modifications that have been introduced to meet the criticisms. In particular he describes the dovetailing of the model with quantum field

125

theory, and the support that the latter offers to the central hypothesis. In all he says, Bohm is appropriately tentative. His weapon against dogmatism is open-mindedness. But, as he points out, it is not enough to expose the weaknesses of current theories; a fresh theory is needed to disturb complacent slumber. If he should succeed in rousing his colleagues, he can expect a formidable assault on his views. Sleepers do not like to be awakened. There is no doubt he can take care of himself.

A final word on his philosophic opinions. Bohm's basic notion is that nature is "qualitatively infinite." This means that the qualities that man has encountered in experience thus far in no sense limit or even foreshadow what he can expect to encounter as he continues to probe, to vary his methods, to flex his imagination. One detects in Bohm's views the strands of many older philosophies — from those of the pre-Socratics to Henri Bergson. The more creative physicists have in recent years cultivated philosophy. They are usually disinclined to admit to this weakness, having in mind David Hume's abjuration to commit to the flames all works of metaphysics as mere sophistry and illusion. But there is no escape, even if it be only to embrace antiphilosophic philosophies. For the physicist has come to realize that if he throws philosophy in the fire, his own subject goes along with it. In this century the professional philosophers have let the physicists get away with murder. It is a safe bet that no other group of scientists could have passed off and gained acceptance for such extraordinary principles as complementarity, nor succeeded in elevating indeterminacy to a universal law. Bohm's challenging book perhaps marks the beginning of a retreat from high-flown obscurantism and a return to common sense in science. I use "common sense" as William Kingdon Clifford used it: science without priestly pretensions or dogmas.

DETERMINISM AND INDETERMINISM

"In asking what we mean by this [cause] we have entered upon an appalling task. The word represented by 'cause' has sixty-four meanings in Plato and forty-eight in Aristotle. These were men who liked to know as near as might be what they meant; but how many meanings it has had in the writings of the myriads of people who have not tried to know what they mean by it will, I hope, never be counted." — From *On the Aims and Instruments of Scientific Thought*, by William Kingdon Clifford

THE "appalling task" has not grown lighter. Clifford's words were spoken to the members of the British Association in 1872; now, eighty-nine years later, neither scientists nor philosophers are in closer agreement

127

among themselves as to the meaning of cause. Is it a hypothesis? A principle? A law? An indispensable crutch of thought? A mere fashion of speaking? How is causality related to determinism? Has modern physics, in undermining the latter notion, subverted the former? The word "cause," said Bertrand Russell, is so inextricably bound up with misleading associations "as to make its complete extrusion from the philosophical vocabulary desirable." A drastic measure; and even if it were justified, one could not expect it to be adopted overnight. Meanwhile thoughtful men continue to revolve the question, for it is held to be crucial not only by philosophers and psychologists but also by those who work at the extreme boundaries of physical science — the theorists concerned with the behavior of very small things and those concerned with the structure of the universe as a whole.

The late Ernst Cassirer reflected on the subject for many years. His philosophic interests were wide and varied. He gained distinction by his writings on Leibniz, Kant, epistemology, symbolic forms, the role of myth in political theory. His best known work, *Substance and Function*, published in 1910, dealt with the philosophy of the exact sciences, and it is to this department, in which he had been thoroughly trained and to which the whole style of his thinking was inclined, that he returned in *Determinism and Indeterminism in Modern Physics*, first published in Sweden in 1936. (It was in that country that he found asylum from the new German order; in 1941 he came to the United States and taught successively at Yale and Columbia. He died in 1945.) It appears now in an excellent English translation by O. T. Benfey of Earlham College, Indiana, with a preface by Prof. Henry Margenau of Yale.*

One might suppose the essay to be out of date. Physical

---

* Ernst Cassirer, *Determinism and Indeterminism in Modern Physics*, New Haven, 1956.

science has moved rapidly in the last two decades, and while the problem of this inquiry has agitated thought since antiquity, it has not been as shaken up in the preceding twenty-five centuries as in the first half of ours. But as Margenau points out, Cassirer's views have not been passed by. For some years we have been living off scientific capital. It consists of the grand theories of Max Planck and Einstein, of the brilliant insights of the 1920s, among them Schrödinger's equation, the uncertainty principle and quantum mechanics. The spectacular advances of nuclear fission, newly discovered particles, new phenomena and facts, are to a large extent the fruits of earlier speculations, a "vast proliferation of the methodological resources contained in the creation of the quantum theory."

Cassirer followed and fully understood the revolution in physics; moreover, his appraisal of its implications was ahead of the day. Some contemporary commentators wrung their hands in despair; some attempted to cover up the cracks in the old foundation with a mortar of words. It was common opinion among physicists that the uncertainty principle imposed fringe limits on very refined measurements, injected certain uncontrollable errors into experiment, but did not force a profound re-evaluation of the meaning of reality. In some quarters the discomfiture of physics was an occasion for jubilation, for with the supposed downfall of determinism free will was restored, and the human spirit no longer had to be regarded as a trolley running in predestinate grooves. Cassirer shared none of these moods, subscribed to none of these opinions. Too much fuss, he felt, was being raised over a rather narrow interpretation of causality, and too little attention given to the conceptual changes that the uncertainty principle required. It made no sense to try to save the classical models by adding curlicues. Such concessions merely invite other concessions. It was no use pretending that everything in physics was as tidy as before except in certain special regions where the broom would not quite fit into the corners. Nothing short of a reform

129

of the concepts of "physical system" and "physical state" was called for. This would show the directions that physics should take, the way to distinguish between meaningful and meaningless questions, the limits inherent in its methods; at the same time it would demonstrate that causal descriptions, as most philosophers have thought of them, can be retained. To Cassirer's mind the new theories of physics promised not chaos but a renaissance.

Cassirer's book surveys the evolution of the notion of cause. He begins with a discussion of the "Laplacean Spirit," which has played an important if not a decisive role in the controversies about causality. Laplace's famous dictum, which appears in his introduction to the *Théorie analytique des probabilités*, asserts that an "all-embracing spirit," which knew all about the universe at a given instant, could by mathematical analysis discover everything that had happened in the past and everything that would happen in the future. For Laplace this was "an effective metaphor" serving to contrast the concepts of probability and certainty. But the underlying theme was gradually transformed into a broad principle of far-reaching significance in the evolution of physical thought.

It was not a new theme. It was in fact the "pregnant summary" of that world view from which sprang the great philosophical systems of the seventeenth century, the systems of classical rationalism.

That everything is brought forth through an established destiny [wrote Gottfried Wilhelm Leibniz] is just as certain as that three times three is nine. For destiny consists in this, that everything is interconnected as in a chain and will as infallibly happen, before it happens, as it infallibly happened when it happened. . . . Mathematics . . . can elucidate such things very nicely, for everything in nature is, as it were, laid off in number, measure and weight or force. If, for example, one sphere meets another sphere in free space and if their sizes and their paths and directions before collision are known, we can then foretell and calculate how they will rebound and what course they will take after the impact. . . . From this one sees then that everything proceeds mathematically — that is, infallibly — in the whole wide world, so that if someone could have a sufficient insight into the inner parts of things, and in addition had remembrance and intelligence enough to consider all the circumstances and to take them into account, he would be a prophet and would see the future in the present as in a mirror.

The clever fellow with the synoptic eye had not yet been born, nor was he likely to be. Still, he embodied an ideal for science as well as philosophy. One might say that if all men were virtuous there would be eternal peace; and though one recognized that all men were not virtuous, and would never be, still the cause of peace would be furthered by encouraging virtue. Similarly, one accepted the view that the universe is a machine, and if its workings could be grasped completely, past and future would be open; and though one recognized that a complete understanding of the universe is impossible, by extending and improving observation one could hope to learn the exact laws of phenomena, the principles by which everything is caused and determined.

It must not be supposed that this formula satisfied the majority of philosophers. Among those who attacked it, especially its treatment of causality, were David Hume and Immanuel Kant. Hume's skepticism led him not only to reject a substantial material world, but to question causality as a universal principle of physics. Our knowledge of the outside world, he said, consists entirely of sensations. We see a moving billiard ball collide with another at rest; the second ball is set into motion; we conclude that the motion of the second ball was *caused* by the first. In truth we know nothing of the actual forces at work; all we know is that there is a regularity in the behavior of our cognitive powers. The causal idea emerges

131

from the fact that imagination and understanding "cannot escape the constraint of association and the force of habit."

Kant also rejected the notion that cause is a simple relation between things. But he departed from Hume in refusing to deprive the causal notion of any other than a psychological base. The mind is orderly, to be sure; but we must not assume that this orderliness prevails only inside the head. There is a constancy in "outer" occurrences, a lawfulness in nature, to which the orderliness of "inner" occurrences is related. "If," said Kant, "cinnabar were sometimes red, sometimes black, sometimes light, sometimes heavy . . . my empirical imagination would never find opportunity when representing red color to bring to mind heavy cinnabar."

The criticisms of Hume and Kant shed a good deal of light on the notions of cause and determinism. Clarification on these matters was a high service to the methodology of physics. What philosophers say is not invariably interesting or essential to physicists. Galileo's and Newton's discoveries, Boyle's and Coulomb's laws and Maxwell's equations needed no metaphysical midwife to be born. But the more elaborate the structure of physics, the more numerous its observations, statements, laws and theories, the deeper the search for unifying principles. This attempt to interpret larger and larger chunks of the world compels physicists to become philosophers and invites philosophers into physics. In the nineteenth century this coalescence of what are in many respects quite distinct ways of thinking is plainly visible.

Field theories of electricity and magnetism forced a radical re-evaluation of the older substance-theory of matter. Statistics introduced into science a new method and a new instrument of description. Thermodynamics led to an important transformation in the world view of physics "in the formal as well as in the material sense." It was no longer possible to rest on a simple "contact" theory of causation. There were philosophers who had rejected this notion long ago; now physicists had to

132

abandon it. Nature was not to be entirely explained in terms of hard objects bumping into each other. The existence of hard objects was doubtful, and even if they existed they appeared to bump into each other from a distance. The concept of continuity was in danger, the jurisdiction of dynamic laws was curtailed; the causal principle itself became the subject of irreverent debate.

Ernst Mach put his finger on an important point, related to the views of Hume and Kant. However one may define the causal principle, he said, it is certainly not a natural law in the usual sense of the words. There is no cause and effect in nature as such "for nature is present only once, and those same events to which we refer when we say that under the same circumstances the same consequences occur do not exist in nature but only in our schematic reproduction of it." Helmholtz expressed himself along similar lines. The framework of physical thought is the principle of orderliness according to law in phenomena. This is the "first product of the conceptual grasp of nature"; cause can be understood and justified only in this sense. Ordinary language uses the notion of cause in a confused way for antecedent or condition: if A regularly precedes B, then A is cause; and likewise, if B never occurs unless A occurs. But orderliness is a much broader and more general concept, and though we cannot be certain a priori that it pervades every corner of the universe, it is an indispensible regulative principle of thought, the only guide to inquiry. If every atom differed from every other, if no regularity were perceivable, "our intellectual activity would necessarily come to rest." The investigator must have faith in a kindlier nature (God, said Einstein, is not malicious). He must give heed, according to Helmholtz, to only one counsel: "Trust the inadequate and act on it; then it will become a fact."

We may ask how statistical laws fit into this philosophical view. In classical physics a distinction was made between dynamic and statistical laws. It was a Leibnizian distinction:

133

the two modes of description were contrasted as "determined" and "undetermined." The behavior of a falling body, described by the laws of motion, was determined; the behavior of a gas, described by the laws of statistical mechanics, was undetermined. What did this distinction mean? For one thing, of course, it implied that some laws were "exact," while others were mere approximations. The fact that observations of instances of dynamic laws never yielded exact results in the mathematical sense was not felt to be a blemish. The apple that fell on Newton's head obeyed his laws exactly; if experimental evidence as to its path and rate of fall exhibited minor variations, this was the experimenter's fault and not due to imperfections in the dynamic laws of motion. Statistical laws, on the other hand, were a makeshift, the best that the physicist confronted with horrendous aggregates could achieve. This did not mean that the individual molecules of a gas do not obey the laws of classical mechanics; it meant that in dealing with a horde of them it was feasible only to treat them statistically.

There was, however, an even more significant aspect to the distinction between "determined" and "undetermined" descriptions. The formulas of Leibniz and Laplace made predictability the criterion of causality. If events are causally connected in a chain, they are determined; since they are determined, they are, at least theoretically, predictable. The laws of motion made possible the prediction of the path of a moving body; in statistical laws, however, the linkages were vague, and one could only prophesy what was probable or likely. But unless it were assumed that *all* phenomena are governed by dynamic laws (though the difficulties of observation made it necessary in some cases to describe what happened statistically), there would in certain parts of nature be a failure of the causal principle. This could not be permitted. Therefore it followed that statistical laws could not be regarded as a complementary method of description, equally as valid as dynamic laws, but only as an ingenious and somewhat inferior substitute for them.

Helmholtz saw more deeply into the problem. His definition of causality was not tied to predictability. He did not belittle the importance of prediction in pure science, in technical mastery of nature, in everyday life. Yet it seemed sheer anthropomorphism to insist that unpredictable events lay outside the causal relation. It might well be there were phenomena that would forever elude exact description, which could be encompassed only by statistical law, but this could not be taken to mean that such phenomena stand above the orderliness of the universe. Max Planck, writing on causality fifty years later, was even more explicit. No doubt, he said, an event is to be regarded as causally determined if it can be predicted with certainty. But this simple definition is inadequate. For although predictability is an infallible criterion for the presence of a causal nexus, we must not infer that predictability is *equivalent* to causality. He pointed out that "even in classical physics it is not possible in a single instance really to predict a physical event accurately." Thus, if predictability is adopted as the strict criterion for causality, we are confronted with the fiasco of being forever unable to verify causality in a single concrete instance.

From this dilemma there is no escape unless we are prepared to define causality as a proposition concerning cognitions, not things and events. In Cassirer's words: "we must think of causality as a guide-line which leads us from cognition to cognition and thus only indirectly from event to event, a proposition which allows us to reduce individual statements to general and universal ones and to represent the former by the latter. Understood in this sense, every genuine causal proposition, every natural law, contains not so much a prediction of future events as a promise of future cognitions."

Nothing that has happened in twentieth-century physics,

says Cassirer, weakens this interpretation of causality. More than half his book is devoted to it.

✓ Though it effected fundamental transformations of the concepts of time and space, the theory of relativity could be assimilated into the classical mode of thought; not so the quantum theory. It exploded the notion of continuity; it created a special body of laws for small things, inapplicable to large things; it dethroned determinacy and made randomness king; it made philosophers take to their beds and physicists flee to shelter. The quantum theory, as Planck aptly remarked, is a "dangerous foreign explosive which has caused a gaping rift throughout the entire structure [of classical physics]."

We start with Planck's assumption, the result of an attempt to explain on thermodynamic principles electrodynamic phenomena and the laws of emission and absorption of light and heat radiations. He introduced the hypothesis of the elementary quantum of action, and the assumption "that the interchange of energy between oscillators takes place only in integral multiples of a definite quantity ε." At this point a sharp break was made with the basic tenet of classical theory, that the energy of a wave spreads continuously in space. Einstein introduced the concept of the light quantum; Niels Bohr carried the theory further and made it the basis of behavior of his model of the atom. The tentative assumption became a theory, the theory a principle; "totally different and apparently heterogeneous groups of phenomena" were brought together under it. A new world picture came into focus.

But the climax was still to come. It arose in the development of the specialized mathematical method for handling the theory — quantum mechanics. The quantum theory envisaged light possessing a dual nature: in some respects it behaved like a wave, in others like a particle. This dualism explained brilliantly many features of the atom, but also gave rise to inconsistencies. In the 1920s, beginning with the work of Louis de Broglie, reforms were instituted to eliminate these inconsisten-

136

cies. The suggestion was made that matter, like light, is both particle and wave, and quantum mechanics was invented as a tool to handle the broadened theory. Erwin Schrödinger took the approach of wave mechanics, Werner Heisenberg that of matrix mechanics. The mathematical differences between these methods do not here concern us; it need only be observed that elements of both theories were combined by P. A. M. Dirac into a new mathematical form, which is the basis of modern quantum mechanics.

Now, quantum mechanics is a statistical discipline. It presents no exact description of an individual particle and makes no exact prediction of its behavior. However, it can make very accurate predictions as to large aggregates of particles, and can show certain correspondences between quantum and classical mechanics. For example, Heisenberg conceived a matrix that corresponds to the momentum or energy of electrons, and combined the matrices according to laws "in the same way the corresponding quantities in classical mechanics are combined by the equations of motion." Quantum mechanics thus has been able to incorporate the laws that were the foundation of classical mechanics.

And yet, as must be evident, the methods of quantum mechanics represent a profound change in outlook, a change that was crystallized by Heisenberg's principle of indeterminacy, enunciated in 1927. All statements in physics, he said, are relative to the means of observation used. This limitation can be disregarded when one experiments with things of ordinary size, but it is decisive in the realm of the very small, just as a grain of sand will not clog the gears of a cement mixer but becomes disastrous in the gears of a watch. Since we are dependent on instruments, there is a limit "not only to our experimental technique but also to the formulation of physical concepts." For a concept not to be "empty or ambiguous," there must be a way of validating it by experiment.

Suppose we speak of the position of an electron. The electron is not in heaven, nor is the concept of "position" an ideal. We must indicate how position is to be determined. Imagine illuminating the electron under an ultramicroscope. We may accept the fact that the shorter the wave length of the light, the more precise will be the determination of position. Unfortunately, however, the shorter the wave length, the more massive the "Compton effect," according to which the light hits the electron and changes its momentum. Therefore it is "basically impossible to measure simultaneously the position and velocity of an electron with any desired accuracy." If we increase the accuracy of one measurement, we decrease the accuracy of the other. "If we designate the uncertainty in the measurement of position as Δx and that in the measurement of momentum as Δmv, then the product of these quantities can never be reduced below a certain value which is of the order of magnitude of the elementary quantum of action."

What are the implications of the uncertainty principle? Does it constitute a denial of causality? Heisenberg said yes. Since there are limits to the accuracy of measurement, there are limits to the predictability of the path of a particle. In his words: "In the precise formulation of the causal law: 'If we have exact knowledge of the present, we can determine the future' not the consequence but the antecedent is wrong. As a matter of principle, we cannot come to know the present in all its determinative factors. . . . Since all experiments are subject to the laws of quantum mechanics (and thus to the equation $\Delta x \Delta mv \gtreqless h$) the invalidity of the causal law is definitely established by quantum mechanics."

Many physicists share this opinion. Some, including Einstein, profess to see in the statistical nature of quantum mechanics a "transitional stage," a "halfway scheme" that needs to be, and perhaps can be supplemented to encompass individual events. Cassirer took a different view. He felt that quantum mechanics is an "ultimate sort of discipline," not perfect, yet representing the finest-meshed net we can hope to throw about physical reality. But he vigorously rejected the idea that acceptance of the uncertainty principle entails the abandonment of causality. Heisenberg's demonstration is leveled, in Cassirer's opinion, against an interpretation of causality that is subject "to grave theoretical faults, even without considering the uncertainty relations." Taking Helmholtz's refined formulation of causality as independent of predictability, he reasoned that uncertainty is not fatal to the causal principle. Indeed, quantum mechanics itself could not exist without making use of "pure and typical causal principles," namely, the classical laws of conservation of energy and momentum.

We must not ask of the notion of cause more than it can give, and we must not misapply it. Planck once pointed out that whether a physical quantity is "observable in principle" or whether a given question is meaningful in physics are not matters to be decided a priori. It depends entirely on the theory we adopt, for the theory shapes our instruments, our methods, our interpretations. We are not simply cameras or mirrors. It may be we can still save the notion of a line of demarcation between the observer and the thing observed; but, if so, at the line itself there is an uncertainty, an uncontrollable ripple, one might say, produced by the interaction of the seer and the seen. The new physics disenchants us as to the firmness and fixedness of substance. "We no longer have absolute, completely determined entities, from which we can immediately read off the laws and to which we can attach them as their attributes." The content of our empirical knowledge consists of no more than the "totality of observations which we group together in definite orders . . . and which we can represent by theoretical laws." Quantum mechanics does not deal with things whose laws we seek to discover; instead, from observation we constitute the things. Atomic physics deals "with the nature and structure not of atoms but of the events which we perceive when observing the atom."

It is against this background that one must view the survival of the causal principle. It can be united with diverse conceptual schemes, but each amalgamation must be effected with care; causality is not a shoehorn that fits every shoe. Cassirer reminds us that causality has had many trials in the history of physics. It was thought to be endangered, for example, when physical thought relinquished the notion of spatial contiguity as the essential condition of cause and effect. When Newton introduced "action at a distance," there was widespread fear that the attempts to explain nature would collapse because of the "occult" qualities ascribed to gravitation. But the causal principle endured; and the exertions to preserve it have deepened our insight.

We now understand that there are in physics, as in other spheres of thought, unaskable, which is to say, meaningless, questions. (e.g., What is the location of an electron when it jumps from one orbit in the atom to another? Is the orbit of its destination already determined at the instant of departure? What happens to radiation if the process is interrupted before the emission of one quantum is completed?) Many intuitively obvious notions have turned out to be untenable, such as that a material corpuscle has an individual identity. What all this means is that some situations "are not empirically definable for us." Nature is much queerer than we can suppose. It does not mean, however, that nature is capricious or that causality is dead. "What quantum mechanics does," said Dirac, "is to try to formulate the underlying laws in such a way that one can determine from them without ambiguity what will happen under any given experimental conditions. It would be useless and meaningless to go more deeply into the relations between waves and particles than is required for this purpose." This is in fact all that the causal principle requires. So long as observable events can be described with precision in mathematical language, the postulate of the "comprehensibility of nature," which the causal principle contains, is fulfilled.

There can be no talk of a final verdict. De Broglie and

David Bohm have attempted what is called a "causal interpretation of the quantum theory"; the Harvard physicist R. P. Feynman has proposed a theory of quantum electrodynamics in which time flows "backwards"; and Max Born and the late Hans Reichenbach have offered new approaches. But it is my impression that most physicists and philosophers are today a little weary of the causality scandal; at least they seem to feel that until another break-through is made, further discussion is not apt to be very fruitful. Cassirer's interpretation now enjoys, as Margenau remarks, greater popularity than when it was first propounded. I am surprised to find that he overlooks entirely the analysis of causality made by William Kingdon Clifford in the 1870s — even more forward-looking than Helmholtz's. And there are certainly weaknesses and unresolved points in the present treatise, examples of jiggery-pokery and paradoxing, which critics will have no difficulty sticking their knives and needles into. But I have no hesitation in recommending this work of a sensitive, sure thinker. Cassirer's study is spacious, subtle and insightful. He has laid out in clear perspective one of the most complex and vexing questions of thought.

REASON AND
CHANCE IN
SCIENTIFIC
DISCOVERY

―――――――――――――――――――

T HE WORLD rewards discovery. The first man
to climb a mountain or discover a planet, a
cure or a natural law is a hero. It is not always certain, to be
sure, that the discoverer made the discovery. In the realm of
ideas especially, the true inventors are apt to be unknown.
Everything of importance, Alfred North Whitehead observed,
has been said before by somebody who did not discover it. But
the hunger for heroes is as great as the hunger for scapegoats;
and as some innocent men are hanged lest murder go unpun-
ished, some undeserving men are rewarded lest discovery remain anonymous.

Scientific discovery is a subject of lively interest. Almost
everyone is curious about the events leading up to a new drug
or mechanical contrivance. Who thought of it? What gave him

the idea? And so on. The circumstances are embroidered and dramatized; they grow into legends. "Suddenly Edison perceived . . ." "Pasteur happened to glance at the cage and all at once he realized . . ." Lately this harmless subject has attracted the notice of politicians. They pursue it more heavily. How, they demand to know, are inventions made? What shall we feed our children to make more Einsteins? Lincoln suggested his generals might profit from drinking the same whiskey as Grant; statesmen will subsidize any diet that promises a crop of Bohrs.

Are there special factors that stimulate invention? Various eminent scientists have divulged their working secrets. One must "think aside," said Claude Bernard. Creativity, said the chemist J. Teeple, is promoted by two warm baths in succession. Others have suggested black coffee, the study of dreams, long walks, a vegetable diet. The pattern is not altogether clear. We must look elsewhere.

It is not generally appreciated that science is a creative activity closely resembling other creative activities. For some reason science is thought to be a special case. No systematic effort is made to discover Rembrandt's formula; no one supposes that the origins of Mozart's inventions can be laid bare. How does their case differ from Newton's or Fermat's? Cultural historians examine the general conditions of creativity — social circumstances, contemporary thought and so on — but they are not concerned with minute analyses of invention. (John Livingston Lowes's masterly detective story about the genesis of Coleridge's poems "The Ancient Mariner" and "Kubla Khan" is almost unique.) It is rightly assumed to be too difficult to trace the intricate weave of creative endeavor in the arts. We readily accept the fact that we cannot explain the genesis of Beethoven's Fifth Symphony; is it easier to explain the genesis of Kepler's laws?

There is a common notion that scientific discovery is sudden, like stumbling upon a treasure. The story of Archimedes

143

running naked from the baths shouting "Eureka!" is the archetype of discovery legends. The truth is, scientific discoveries are never sudden. Ideas have ancestors; ideas come in families. The culmination of a family of ideas is, of course, easily identified. Here is Gregor Mendel's paper of such and such date; here is the first edition of *Madame Bovary*. But what led to this result? How many generations of thought? What traditions? What errors and false starts? What part was played by chance, reason, experience? We are scarcely better able to answer these questions for the laws of genetics than for Flaubert's masterpiece.

Still, there is a strong urge to try. René Taton's little book, published in Paris in 1955 and later translated (at times quite awkwardly) into English, is the most recent attempt along these lines.* It has the merit of being a modest work. There have, he points out, been many studies concerned with the origins, conditions and character of scientific discovery. The authors of these studies have striven for general conclusions, but Taton considers this "dangerous territory." Instead, he has restricted himself to "a description of the different realms of scientific discovery, its principal factors and its essential aspects, with examples drawn from the various fields." As an unpretentious catalogue, with some interesting entries, his book is useful. It helps us to understand how subtle and many-sided is the process of discovery and how inadequate are the pronouncements purporting to schematize it.

Of the different forms of scientific creativity, mathematical invention has had perhaps the most serious attention. French scientists, in particular, have concentrated on the subject. In 1905 in *L'Intermédiaire des Mathématiciens* the question was raised: Do "mathematical dreams" promote the solution of problems "vaguely studied" in the waking state? The replies were not enlightening. Most of the correspondents had enjoyed

---

* René Taton, *Reason and Chance in Scientific Discovery*, New York, 1957.

neither algebraic nor geometric dreams; the few who were so fortunate emphasized that a solution appeared at the very moment of waking. A much more ambitious enquiry was undertaken by *L'Enseignement mathématique*. Elaborate questionnaires were sent out, and the answers were carefully analyzed by a group of mathematicians and psychologists. Unfortunately the majority of first-rate mathematicians who received questionnaires were unwilling to take the time to reply in detail, and the bulk of the answers came from less gifted research workers. Still, it would be interesting to know what they said, although Taton doesn't tell us.

In 1908 Henri Poincaré gave his famous lecture "L'Invention mathématique." This essay, read to the General Institute of Psychology of Paris, is so well known that I scarcely feel the need for an extended comment. Poincaré speaks of the nature of mathematical aptitude. It cannot, in his view, be reduced to a fine memory or extraordinary powers of concentration. Mathematicians, he says, are rarely good calculators or strong chess players. He confesses that he himself is incapable of doing a sum without making a mistake; and over the chess board, after examining the various possible consequences of several moves, he would end by making the very first move that had occurred to him, forgetting that its obvious dangers had led him to reject it and to enter upon his calculations. In place of memory, the mathematician relies upon a certain "intuition." With this to preside over his thoughts, to carry him along, he need not be concerned over mislaying bits and pieces of the argument. They will reappear when needed, and meanwhile the forward march of the demonstration will be guided by the "intuition of a mathematical order which enables [him] to guess the hidden harmonies and relations."

There follows his definition of mathematical discovery: "What, in fact, is mathematical discovery? It does not consist in making new combinations with mathematical entities already known. That can be done by anyone and the combinations that could be so formed would be infinite in number and the greater part of them would be absolutely devoid of interest. Discovery consists precisely in not constructing useless combinations but in constructing those that are useful, which are an infinitely small minority. Discovery is discernment, selection."

This is a celebrated definition, but I see no reason to celebrate it. It is not trenchant; it does not light up the dark. It says that discovery is hard, originality rare, and a clever man can tell a good thing when he sees it.

Nor does the rest of the essay — though it is a vivid and wonderful fragment of autobiography — carry us much farther along. We have Poincaré's dramatic account of how he discovered the Fuchsian functions ("As I put my foot on the step, the idea came to me"; "As I was walking on the cliff, the idea came to me," etc.), which ascribes inspiration to the working of the unconscious. After days of apparently unfruitful work, and periods of rest, "the unconscious arranges the results of previous periods of work of which the conscious mind is no longer aware." Under the influence of "some aesthetic sensibility" the break-through is made; thereafter, "the mind must implement the inspiration, deduce and order its immediate consequences, arrange a proof, and above all verify the results." Does this explanation get to the heart of anything?

Other mathematicians have even less to reveal. It has always seemed to me that Jacques Hadamard's much-praised *Essay on the Psychology of Invention in the Mathematical Field* is a tedious book, filled with platitudes and irrelevancies. Error, he tells us, plays a part in some discoveries. He also says that there are different kinds of mathematical aptitude; that there are intuitive and logical minds; that precision, "mental order" and "mental pictures" are useful; that some mathematicians look for applications of their theories or are spurred by physical problems, while others are indifferent in this regard.

146

There is no discernible pattern in the history of mathematical invention, at least none shown by Hadamard.

Suppose we take another tack. Is it easier to identify the conditions of discovery in the experimental sciences than in mathematics or other theoretical discourses? Claude Bernard described the stages in the making of a scientific discovery: the scientist discovers a fact; an idea connected with this fact suggests itself; this leads to further experiments; new phenomena appear, and so forth. Bernard was a great biologist, but this pronouncement is insipid.

It cannot be denied that there have been many systematic discoveries. In such cases it has often happened that the problem itself was plain, that no sudden illuminations were required to indicate a fruitful path of inquiry. The right instruments, ingenious method, thorough analysis, systematic exploitation of previous work — each contributed to the successful result. An excellent example is the discovery of the planets Uranus, Neptune and Pluto.

Five planets visible without instruments were known since antiquity. The invention of the telescope in the seventeenth century immediately enlarged the picture of the solar system. Galileo identified the satellites of Jupiter, and a few decades later Huygens discovered Titan, the largest satellite of Saturn. More than a century passed, however, before the list of planets was augmented. In 1781, with the help of a powerful new telescope that he had constructed, William Herschel methodically swept the different regions of the sky. Near the constellation of Gemini he noticed what he first supposed to be a comet. But on the basis of calculations by Laplace, Bochart de Saron, Joseph de Lalande and others, Herschel came to realize that he had found a new planet whose orbit lay beyond Saturn's. The path of this planet — now called Uranus — proved to be very hard to predict. Despite repeated observations and considerable improvement in computational meth-

147

ods, prediction did not agree with reality. Alexis Bouvard therefore advanced the hypothesis that a still unknown planet perturbed the motion of Uranus. Other astronomers were skeptical: it seemed indecent to keep jiggling and rearranging the order of the universe. But in time the hypothesis was accepted.

The next step was to try to confirm it. Several attempts were made, but they involved such complex calculations that the interested astronomers gave up before making substantial headway. In 1841, however, John Couch Adams, a twenty-two-year-old student of St. John's College, Cambridge, tackled and solved this enormously difficult problem. On the basis of refined observations of Uranus, he was able to make a much closer estimate than had yet been achieved of the mass, position and trajectory of the hypothetical planet. Eager with his news, he called upon the director of Greenwich Observatory. There ensued a tragicomedy. It was not permitted to disturb the director, Sir George Airy, at his dinner, so Adams was obliged to leave his card and a summary for the great man to peruse at his leisure. When Airy got around to the task, he was not impressed with the result and did little to follow up. Meanwhile, a young French astronomer, Urbain Leverrier, had — quite independently of Adams — described the location of the new trans-Uranian planet. The data Leverrier had to rely on were highly inaccurate. He had to construct the orbit of the as yet unseen planet from its effect on the motion of Uranus; and this effect was of an order of magnitude "never greater than that of the errors of observation." Yet such was his virtuosity — and boldness — in carrying out this involved and delicate calculation that he was able to send a remarkably accurate set of directions for finding the planet to the German astronomer Johann Galle. Galle received the letter at his observatory in Berlin on December 23, 1846; that very evening he discovered the planet we now call Neptune only fifty-two seconds of arc away from the position Leverrier

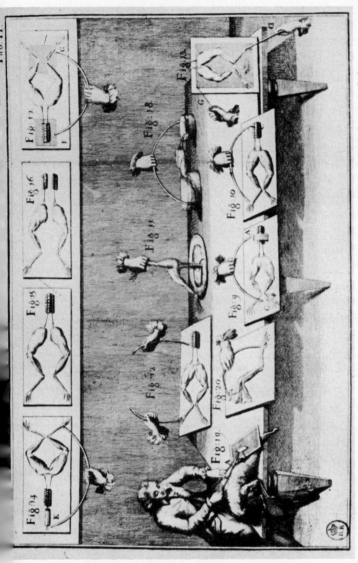

One of the plates from Galvani's work on animal electricity, De Viribus Electricitatis, Utinae, 1792, plate II. The different pictures illustrate some of Galvani's explanations of the effects of discharges from electric condensers (Leyden jars) on the nerves of frogs and other animals. Although Galvani failed to interpret the phenomenon correctly, his work enabled Volta to make the major discovery of the electric cell. (From René Taton, Reason and Chance in Scientific Discovery, by permission of Philosophical Library)

had suggested. "The planet of which you have given the position," Galle wrote him, "really exists."

In announcing his discovery to the Academy of Sciences, Leverrier expressed the hope that continued observation of Neptune would lead to the discovery of another planet more distant from the sun, and this in turn would afford a basis for further discoveries. This expectation had partial confirmation. The American astronomers Percival Lowell and E. C. Pickering, building on the work of Adams and Leverrier, and with the help of powerful and sensitive modern instruments (notably the blink microscope), succeeded in forecasting the position of Pluto. Thus a sequence of plainly linked discoveries stretches from the seventeenth century to the present day. Each has its special features, yet common to all is patient and methodical research. There are no flashes of genius, no sudden inspirations; but the discoveries are on that account neither less dramatic nor less important.

Chance "happeneth to all," but rarely with profit. Luigi Galvani observed the twitching of a frog's leg when an electric spark was produced in the neighborhood of his specimen. In this curiously indirect fashion was discovered the electric current. But Galvani, though he repeated the experiment in different ways, interpreted it incorrectly, and it was Alessandro Volta who gave an explanation that led to his invention in 1800 of the voltaic battery. Galvani, in other words, had the scientific imagination not to disregard what he saw, though he could not fathom its meaning; and Volta, a gifted physicist, followed up the observation and transformed it into scientific knowledge.

Another accidental discovery was made by the French mathematician Étienne Malus, who, while looking from his house at the sunlit windows of the Luxembourg Palace through a double-refracting crystal of Iceland spar, observed to his surprise that when he turned the crystal about its axis, each of the two images would vanish in turn. He realized that the phenomenon was somehow connected with the reflection of light by the windows, and from this inference deduced the theory of polarization by reflection. (The phenomenon was not fully explained until some years later, when Augustin Fresnel and Thomas Young broached their theory of the transverse nature of light waves.) Taton points out that Malus's fortuitous observation "could not have fallen on a mind better prepared to draw the consequences." Undoubtedly the phenomenon had been observed before, but it took a physicist interested in geometrical optics to appreciate its significance.

Alexander Fleming's discovery of penicillin makes an intriguing story. One day in September 1928, while studying mutation in some colonies of staphylococci, he noticed that a Petri dish had been contaminated by a microorganism from the air outside. This was not an uncommon accident, but Fleming was unwilling to dismiss it. On examining the contaminated dish in greater detail, he found that a fungus had attacked the staphylococci and had made a large region of them transparent. He concluded at once that an antibacterial substance produced by the fungi had destroyed the staphylococci. Now, it is important to recall that fifty years earlier John Tyndall, Louis Pasteur and Jules Joubert had made similar observations; but their discoveries (e.g., the air, Tyndall wrote to Huxley, contains bacteria germs "in myriads"), while contributing to the foundations of modern bacteriology, lay relatively unattended for years. Fleming's great advance consisted of a brilliant demonstration of the selective properties of the substance secreted by the fungus and of its action on different species of bacteria.

These results were published in a classic paper, which appeared only eight months after his initial observations. The reception it got was restrained; physicians are not a conspicuously imaginative breed. It is true that Fleming himself did not fully realize the power of the substance he had discovered

151

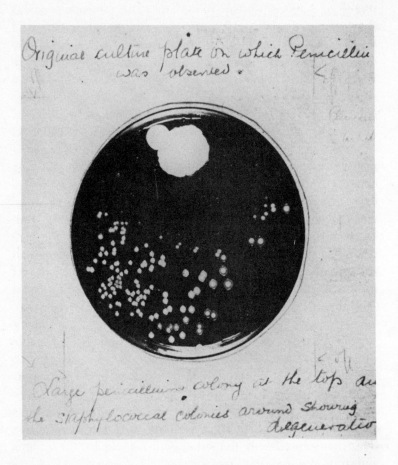

The original plate containing the culture in which Sir Alexander Fleming
discovered the antibiotic action of penicillin. A large colony of Penicillum
can be seen in the upper part of the preparation. The surrounding colonies
of staphylococci show obvious signs of degeneration. (From René Taton,
Reason and Chance. in Scientific Discovery, by permission of Philosophical
Library)

152

and its potentially revolutionary consequences. It took several years of concentrated laboratory work by various bacteriologists and biochemists to master the difficulties of preparing even small quantities of a pure and stable substance incorporating the active principle of penicillin. Finally, as a by-product of the outbreak of the Second World War, the developmental phase was brought to a successful climax. Urgent medical needs led to a massive assault by teams of United States and British scientists and technicians on the problems of large-scale industrial manufacture of the new antibiotic.

For the student of scientific discovery, these circumstances comprise a most illuminating case history. Consider how many elements had to be brought together: a number of precursor experiments by physicists and chemists; an accident caused by inadequate precautions in the laboratory; a talented and perceptive biologist who insisted on worrying an apparently unimportant phenomenon until he had shaken loose its meaning; a follow-up multidisciplinary program of research; the support of industry and government; lavish public expenditures. All these elements went into the complex process of "the discovery of penicillin."

Some discoveries are forgotten for many years (Gregor Mendel's famous researches in heredity are the most obvious example); some are narrowly missed. The little less, as Browning wrote, is "worlds away." Henri Poincaré and Hendrik Lorentz, among others, "approached" the theory of relativity but lacked, in Taton's words, "the courage to make their thoughts explicit." Louis de Broglie ascribes Poincaré's failure to the fact he was a pure mathematician with a "somewhat skeptical attitude towards physical theories." In his view such theories were not "true" but merely "convenient." He was therefore loath to plump for this or that physical model, since any number of others logically equivalent to it were

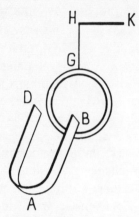

Sketch of apparatus with which Ampère "missed" the discovery of induction. (From Correspondance du Grand Ampère, *Volume Two, page 761)*

equally valid. Einstein's great achievement is, of course, in no way diminished by these circumstances.

Blaise Pascal in his studies of the roulette problem invented a method which opened the way for Leibnitz's invention of the infinitesimal calculus, but Pascal himself regarded his method as no more than an aid to calculation. He had fame enough.

We come now to André Marie Ampère's strange failure to discover induction. Hans Christian Oersted's famous experiments in 1819 had shown the magnetic effects of an electric current. Physicists at once began to search for a reciprocal phenomenon — the electrical effects of a magnet. The first attempts to demonstrate induction failed, for they were based on the mistaken notion that if a magnet were merely placed close to a wire, something interesting would happen. A magnet at rest relative to a conductor produces no current; indeed, if this were not so, the principle of conservation of energy would fail. It remained for Michael Faraday a decade later to show that, for a current to flow, magnet and wire have to be in relative motion; the accepted explanation is that the current depends on the cutting of the magnetic lines of force.

154

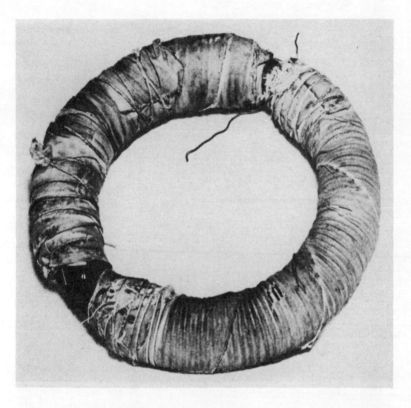

Soft iron ring, used by Faraday in his discovery of Electromagnetic Induction on August 29, 1831. One of the two wires surrounding the core is connected to a battery, the other to a simple galvanometer consisting of a copper wire running over a magnetized needle. On making the primary circuit, the needle is sharply deflected to return to its original position after some oscillations. The needle is also deflected when the circuit is broken. The deflections are due to the fact that a current is induced in the secondary winding whenever there is a change of current in the primary circuit. Despite its primitive nature, this apparatus led to a discovery the importance of which can hardly be overestimated. (From René Taton, Reason and Chance in Scientific Discovery, *by permission of Philosophical Library)*

155

Detail in a plate by Francesco Stelluti (1577–1640), from the Apiarum *of Federico Cesi (1585–1630). This work contains the oldest-known drawings seen through a microscope. (From René Taton,* Reason and Chance in Scientific Discovery, *by permission of Philosophical Library)*

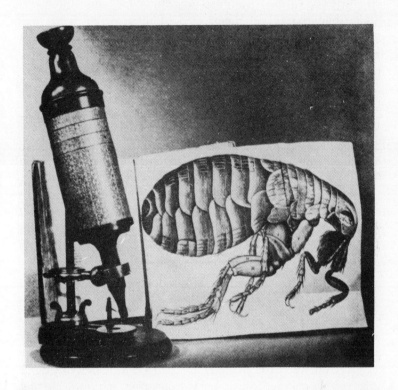

Model of Robert Hooke's microscope in front of his Micrographia *(London, 1655), opened at plate XXXIV, a drawing of the flea. The body of Hooke's microscope was made up of four cardboard tubes fitting into one another. The objective, a biconvex lens of very small focal length, was held in an externally threaded wooden holder by means of a very small diaphragm. The apparatus was focused by turning the thread in a ring attached to the support. The eyepiece, consisting of a planoconvex and a small biconvex lens, was mounted on another wooden holder, fitting over the first of the cardboard tubes. Magnification was of the order of X 30.*

The Micrographia *is a beautiful folio, containing 60 microscopic and three telescopic drawings. The drawing of the flea is very clear, and the caption reads: "The strength and beauty of this small creature, had it no other relation at all to man, would deserve a description." (From René Taton,* Reason and Chance in Scientific Discovery, *by permission of Philosophical Library)*

157

Meanwhile, in the course of other experiments with a different purpose, various instances of induction came into view, but no one recognized them. One of these experiments was performed by Ampère and his friend Auguste de la Rive. A ring made of a thin strip of copper was suspended by a silk thread over a coil of wire wound parallel to the ring. The ring was then placed in the field of a powerful permanent magnet. When a current was set up in the coil, the ring was displaced; when the current was shut off, the ring returned to its original position. What did this mean? The copper ring was a conductor; the turning on and off of the current in the coil produced a magnetic flux, which induced a current in the ring, causing it to move in its magnetic field. This was no surprise to Ampère, who had already demonstrated the attractions and repulsions produced by magnetic forces between adjacent wires carrying currents in opposite directions. But while he recognized what he called the "production of currents by influence" (i.e., induction), the essential idea of relative motion as a necessary condition eluded his understanding. He simply inferred that since the ring remained in a displaced position while the current flowed in the primary circuit (the coil), the induced current in the ring kept flowing during the same period; and when the primary current ceased, the ring went back to its original position because the induced current ceased. Because of this erroneous inference Ampère missed the discovery of electromagnetic induction.

When Faraday announced his own achievement, Ampère reproached himself for having failed to draw the logical consequences of his work. Having, with De la Rive, been the first to obtain a current by induction, he neglected to exploit this epochal feat. Why? His explanation is only partially convincing. "I assure you," he wrote Faraday in 1833, "that at the time I never once tried to find out in which sense a current is produced by induction. I had but one aim in making these experiments, and by taking a look at what I published at the

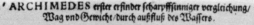

A sixteenth-century etching, illustrating Plutarch's account of Archimedes'
discovery of the law of the upthrust of liquids. Near Archimedes' bathtub
can be seen Hieron's famous crown, and also other objects used in hydro-
statics (spheres of different diameters, tanks with taps at different levels,
etc.). This etching is from Der furnembsten notwendigsten der gantzen
Architektur angehorigen mathematischen und mechanischen Kunst, *W. H.*
Ryff, Nuremburg, 1547. (From René Taton, Reason and Chance in Scientific
Discovery, *by permission of Philosophical Library)*

time, where I described the apparatus that I used, you will
see that I was only concerned with solving the question whether
electric currents are due to magnetic attraction and repulsion
present before magnetization in the molecules of iron, steel
and two other metals, in a state which does not allow them to
exercise any action outside, or whether they are produced at
the moment of magnetization by the influence of neighboring
currents."

That Faraday succeeded where not only Ampère but many
others failed is a typical puzzle of scientific discovery. Taton
does not pretend to pierce it.

159

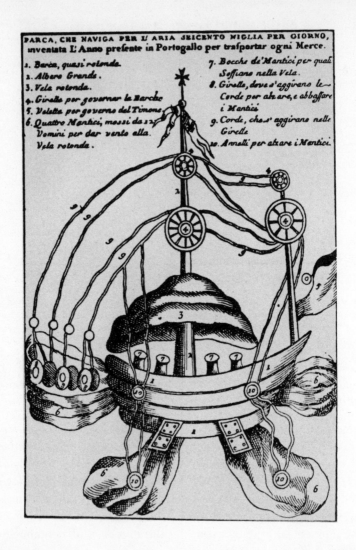

Fantastic design of a "flying boat," attributed to the Portuguese inventor Lourenço de Gusmão. This rare Italian etching, dating from about 1710, is taken from La Machine Volante de Gusmão, d'après une figure comique, *by J. Duhem (Thalès, volume three, 1936, pages 55–67). "A boat, travelling through the air at 600 miles per day, invented this year in Portugal for transporting any kind of merchandise whatsoever. (1)Boat, almost round; (2)*

160

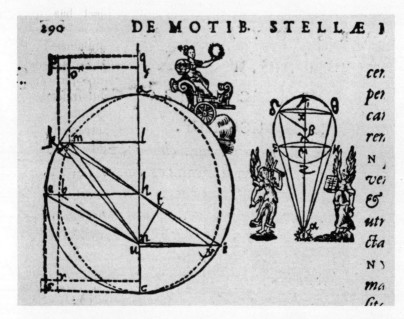

Figure from Kepler's Astronomia nova . . . de motibus Stellae Martis, *Prague, 1609, page 290, where he gives his proof of the ellipticity of the trajectory of Mars. The very clear and convincing drawings are surrounded by allegorical figures. (From René Taton,* Reason and Chance in Scientific Discovery, *by permission of Philosophical Library)*

Top-mast; *(3) Round sail; (4) Pulleys for controlling boat; (5) Small sail controlling rudder; (6) Four bellows for blowing into round sail, operated by ten men; (7) Vents through which the bellows blow air; (8) Pulleys with halyards for raising and lowering the bellows; (9) Halyards; (10) Rings for raising bellows." (From René Taton,* Reason and Chance in Scientific Discovery, *by permission of Philosophical Library)*

The serious student's outlook on the subject has un-doubtedly become more mature. The realization is growing that science, like other creative work, is strongly influenced by social, economic and political circumstances; that the general style of contemporary thought in religion, philosophy and other spheres profoundly affects scientific thought. Indeed, it may be said that these notions are already commonplace. But the nature of the relationships is only dimly perceived and requires much study. It is also increasingly recognized that the acquisition of knowledge is a co-operative task. This does not mean, I should emphasize, that the best research has to be expensive or that progress in tackling complex problems is assured when everyone gets into the act. It is still possible to make important discoveries in the head, and even to make them in one head.

Some years ago Lord Rutherford concluded a lecture on "The Development of the Theory of Atomic Structure" with these words:

I have tried to give you a general idea of the way in which we started to investigate these matters forty years ago, and of the way in which the ideas have developed stage by stage. I have also tried to show you that it is not in the nature of things for any one man to make a sudden violent discovery; science goes step by step, and every man depends on the work of his predecessors. When you hear of a sudden unexpected discovery — a bolt from the blue as it were — you can always be sure that it has grown up by the influence of one man on another, and it is this mutual influence which makes the enormous possibility of scientific advance. Scientists are not dependent on the ideas of a single man, but on the combined wisdom of thousands of men, all thinking the same problem, and each doing his little bit to add to the great structure of knowledge which is gradually being erected.

162

PART
3

BLACKSTONE FOR
THE PROLETARIAT

ANDREI VYSHINSKY'S *The Law of the Soviet
State** is a large, serious tome of such di-
verse content that it bears some resemblance to Mrs. Murphy's
famous chowder, even to the unmistakable presence of a pair
of toiler's overalls. Having diligently read through its 700
pages, I cannot in my attitude be accused of flippant super-
ficiality; nor do I wish to be understood as implying by my
comparison that there is not much of use and solid value to be
found in Vyshinsky's wide-angle survey. With a group of
anonymous collaborators he has assembled an extraordinarily
variegated, highly seasoned book that will satisfy and/or irri-

* Andrei Vyshinsky, *The Law of the Soviet State*, New York, 1948.

tate the reader, depending on what he expects to find in a volume bearing this title.

The Law of the Soviet State is the standard reference text for Soviet students of law and government. It is also, as Professor John N. Hazard points out in his excellent introduction, a "militant handbook" for Soviet officials, a point to which I shall return. Vyshinsky's contribution is apparently much greater than that of the average editor to a work of this kind. This is understandable on considering that the Soviet Union is a place where consequences are apt to follow from what one publishes. Vyshinsky's active participation is indicated by the energy, aggressiveness, vehemence and invective that characterize much of the writing. His reputation as a diplomat who speaks his mind even when other diplomats are present, and as a prosecutor who approaches almost every issue as though it was a criminal to be exposed rather than a problem to be elucidated, is fully sustained.

The setting in which the book was written demands some notice. It appeared in 1938, two years after the adoption of the new Soviet constitution. This long-awaited event seemed to promise, Hazard writes, an end to many temporary, discriminatory, revolutionary controls and the beginning of an "age of greater tolerance" in the social and political sphere. But in the period 1936–1938 conditions in Europe deteriorated rápidly; the Second World War seemed inevitable, and Soviet policy, both foreign and domestic, bent to the storm. Nothing (certainly no doctrinal preconceptions) could be permitted to impede the U.S.S.R.'s preparations for the approaching conflict. Several of the freedoms promised by the Stalin constitution never came alive (it is idle to speculate whether they might have been put into effect had there been no military threat); and programs designed to bring nearer the goal of communism (when the state, its machinery having become superfluous, would "wither away") were discarded.

For this utopian doctrine of gradual atrophy, Vyshinsky, I

gather, never had much use. For one thing, Lenin had made it clear that "Marxism provided no precise blueprint for the future government of revolutionary Russia." Step-by-step experimentation would be required, he thought, to evolve the forms best suited to existing conditions and forces. Furthermore, Vyshinsky felt that the immensely difficult and precarious task of achieving stability while in rapid motion, of gaining the benefits of permanence while in transition, of establishing a viable order without extinguishing the precious spirit of revolution, could only be carried through by a variety of traditional "legal means," wielded by the proletarian dictatorship: this, in turn, presupposed a strong central state rather than a weakling confederation working toward its own dissolution. That opinion underlies Vyshinsky's entire conception of jurisprudence, and influences the treatment of every major problem raised in his book.

Vyshinsky's survey scans the genesis and meaning of the constitution, the social organization of the U.S.S.R., its economic and administrative system, its body of public law, its courts and legislative units, the rights and obligations of individual citizens and of member republics. It contains a mass of economic data, a wholly inadequate description of the role of the Communist Party, and very little by way of substantive detail as to how the machinery of state actually works as compared with how it is supposed to work.

I cannot hope in this space to explain — even if I were qualified to do so — the ponderous, complex issues of economic, political and legal theory raised in Vyshinsky's analysis and rationalization of developments in the structure of the Soviet state. A few examples will serve to illustrate his didactic method and may furnish glimpses of Soviet law and thought.

1. The great socialist principle, "From each according to his capacity, to each according to his toil," is, as Vyshinsky points out, "unjust" from the viewpoint of communism. Com-

167

munist doctrine, in its purest form, would require instead, "From each according to his capacity, to each according to his need." But this is frankly recognized as unattainable because "it is impossible (without falling into utopianism) to suppose that, after overthrowing capitalism, people will learn at one stroke to work for society *without rules of law of any sort*, and the abolition of capitalism *does not at a stroke give* economic promises of *such change*." (His italics.) This is a typical instance of Vyshinsky's practical approach to matters involving Marxist-Leninist theories. He who works eats; he who doesn't doesn't; he (or she) who works exceptionally well as a riveter, chess player, geneticist (or child bearer) is entitled to "rewards, prizes and exemptions." Nothing in this doctrine, I take it, to alarm General Motors.

2. Bourgeois democracies move under the principle of the "separation of powers," thus preventing abuses that might result from vesting too much authority in a single person or branch of government. The U.S.S.R., says Vyshinsky, does not need this bourgeois principle of checks and balances:

From top to bottom the Soviet social order is penetrated by the single general spirit of the oneness of the authority of the toilers. The program of the All-Union Communist Party (of Bolsheviks) rejects the bourgeois principle of separation of powers. The unity of the authority of the toilers, embodied in the highest organs of that authority, expresses their democratic nature and the sovereignty of the Soviet people.

On the other hand this should not lead to the belief that there is in fact no separation of powers in the U.S.S.R. administrative machinery, for according to the very next paragraph:

In the U.S.S.R., authority has its beginning in the genuine popular sovereignty personified by the Supreme Soviet of the U.S.S.R. This

168

is not incompatible with limiting the jurisdiction of authority as between separate organs. *Such limitation flows out of the extraordinarily complex functions of the Soviet state machinery governing both people and economy.* (My italics.)

This explanation, if I understand it correctly, might be paraphrased as follows: In bourgeois countries automobiles are made with brakes. In the U.S.S.R. this device is rejected as a deliberate attempt to exploit the drivers. Soviet automobiles have no brakes. This does not mean that the Soviets have no way of stopping their cars. Indeed, every car stops itself due to its complex machinery and the friction between the several parts.

3. "Soviet democracy and the Soviet state are a million times more democratic than the most democratic bourgeois republic." (I can think of no other nation, except the United States, capable of saying foolish things so earnestly.) The fact is borne out by the Stalin constitution, which assures "the right to labor; the right to rest; the right to material security in old age; . . . the right to education; freedom of conscience; truly democratic freedoms of speech, of assembly, of the press, and of meetings; and the right to be united in social organizations."

Occasionally, a few restraints are necessary. Thus: "In our state, naturally, there is and can be no place for freedom of speech, press and so on, for the foes of socialism." The granting of freedoms of any kind to those unworthy of freedom — presumably those engaged in un-Russian activities — must be classified as a "counter-revolutionary crime." Freedom of assembly, of meetings, parades, demonstrations and so on "are the property of all the citizens in the U.S.S.R., fully guaranteed by the state *upon the single condition that they be utilized in accord with the interests of the toilers and to the end of strengthening the socialist order.*" (My italics.)

The right to unite in social organizations is absolute; but the

169

Soviet state "naturally does not include freedom of political parties . . . inasmuch as this freedom, in the conditions prevailing in the U.S.S.R., where the toilers have complete faith in the Communist Party, is necessary only for agents of fascism and foreign reconnaissance whose purpose is to take all freedoms away from the U.S.S.R. and to put the yoke of capitalism upon them once more."

The secret police in Russia has a long and unsavory history. Fortunately the new constitution guarantees the citizen the inviolability of his person against invasions by the Soviet FBI. Thus, by example, "No one may familiarize himself with a citizen's personal correspondence, with one exception only: in case this is necessary to disclose crime and detect a criminal."

4. Frequently in the midst of a serious juridical exposition Vyshinsky launches into a stirring panegyric on strangely peripheral topics: the beauties of Moscow; Soviet boundaries, which are "sharply guarded by our heroic border troops . . . [and] four mighty Soviet fleets"; Marx, Lenin and Stalin; the Soviet coat of arms; and so on. For instance: "On the coat of arms of capitalist states birds of prey and beasts — eagles, hawks and lions — are most widely depicted, characterizing the rapacious policy of the states' dominant class." The proletarian dictatorship has for its coat of arms the familiar emblem of toil, the hammer and sickle; underneath are the words: "Proletarians of all countries, unite!" Above is the star of the Red Army — "a menacing reminder."

As characteristic as these examples may be, they give a false impression of the whole book, an impression valid certainly on only one level. For it is important to remember that *The Law of the Soviet State* is written on several levels, with several audiences in view. This accounts, I believe, for many of its puzzling, contradictory features: ideas concealed within ideas; crude arguments and subtle ones; bluster and sobriety; honesty and exaggeration; Slavic realism and Teutonic metaphysics; sensible exposition interlarded with boisterous idio-

170

cies. Vyshinsky's book starts out as a straightforward textbook on constitutional law. The descriptions of the legislative system, of the various commissariats, of the function of the courts, of the formation of the constitution are orthodox and informative. (The style of presentation is, to be sure, incredible. Hugh W. Babb, the translator, seems to have struggled manfully with his immense pile of prose, but sentence after sentence resembles a Chekhov or Gogol caricature of the literary style of a Russian bureaucrat.)

Second, it is a revolutionary handbook, a guidebook and a practical memorial whereby men may be instructed and at the same time warm themselves, in De Tocqueville's words, at the fire of their fathers' passions. Third, it is a hagiography, a tribute to various saints: Marx, Engels, Lenin, Stalin. Fourth, it is a polemic against sundry "wreckers," "bandits" and "gangsters" (Trotsky, Bukharin, etc.). Fifth, it is a declaration of piety and zealous conformity by its authors (Vyshinsky and collaborators). Sixth, it is a show for foreigners, an exhibition of the imposing structure and legal system of the U.S.S.R., a refutation of the "slanders of enemies."

Finally, the book shows evidence of being in conflict with itself on major doctrinal issues. *Destruam et aedificabo* ("I shall destroy and I shall build") were the words appearing on one of Proudhon's books. We may, in a sense, regard this not only as Marx's motto (as Isaiah Berlin points out) but as the motto — and the dilemma — of Vyshinsky as well. How, side by side, to promote obedience and to encourage rebellion, to destroy tradition and yet to build with the aid of its forms? These problems beset the teacher, and baffle him, no less than they baffle and beset the politician.

Vyshinsky's treatise is invaluable for ignorant foreigners like us who live behind the Iron Curtain — for the curtain, of course, has two sides and we are as much behind one as the Russians are behind the other. Each of several chapters contains more information about the Soviet system than is to be

171

found in any year's ration of foreign dispatches, or in the detailed, authoritative treatises on Russia written by those who left there in the days of Kerensky, or in the great collection of confessions, reminiscences and apologias by reformed Marxists, unreformed Czarists, renegades, police spies, informers and other disinterested observers who have jumped, crawled, run, flown, swum and burrowed to freedom. Vyshinsky makes the Soviet state out to be believable and the framers of its laws neither imbeciles, maniacs nor beasts. Undoubtedly they differ from our own politicians, though not as much as you might think. Reading this book is not conducive either to hating or loving the Russians. Neither emotion is desirable, much less required. It should, however, help to a saner understanding of how the Russians differ from us politically. This is required, absolutely, for policy makers as well as plain men.

PATTERNS
OF PANIC

THE WORD "panic" is derived from the name of a disagreeable deity of ancient times who watched over shepherds. Besides a pair of horns, a flat nose, and the legs and feet of a goat, Pan possessed a furious temper, which when roused made even the gods flee. It is not held that Pan actually inflicted physical violence on his victims; it served his purpose either to demoralize them by brutish outbursts or, more subtly, to infect them with uneasiness and vague apprehensions.

Panic is defined as unreasoning fear, a sudden seizure of terror, usually out of proportion to the cause. For the individual it may mean headlong flight, running amok, or some

173

less aggressive hasty action; in the group, it becomes a "collective explosion," as Dr. Meerloo* describes it, and often culminates in mob fury, riot, lynching, stampede.

Panic, however, need not be overt; it is a complex reaction assuming many different and even obscure forms. One person may be routed by noise or the fearful reality of battle, another by the sight of fire, blood, pigeons, snakes, drunkenness; yet another may succumb to a shadow, a fantasy, perplexity, "melancholy," "rapture," or some purely inner conflict; the frustration of expectation also arouses such strong fear and anxiety that actual danger may be preferred to the anticipation of it. The ability to endure anxiety and uncertainty is highly variable; some persons placed under stress crack up immediately; others can carry the burden for long periods; others again — the phenomenon of post-battle shell shock is familiar — surrender to terror when every obvious cause for alarm has been removed.

There is latent panic in all of us. Man is especially vulnerable because his natural biological defenses are poor. Compare him in this respect with a crocodile, an eagle, a lion or a clam. Primitive man, says Meerloo, "lived in a chronic panic, a chronic alert, a chronic mobilization for flight." This universal on the subject of Pleistocene psychoses sounds to me a little glib. We can, however, accept with confidence his assertion that man is not adjusted to his environment as are animals (even our immediate ancestors were masterful climbers, an essential defense in the struggle of life), and that he must substitute for fang and claw tools and superior reasoning power. Obviously in striving for mastery over nature his intelligence has served him well; less well — though even in this domain it has not altogether failed him — in liberation from the congeries of irrational fears. In panic man's best weapon is aban-

* Joost A. M. Meerloo, *Patterns of Panic*, New York, 1950.

174

doned, and he is left the more naked, as he is the more civilized, to face danger, real or imagined, with nothing better than his vestigial talents for flight.

Dr. Meerloo's book is less concerned with "solo-panic" than with collective psychological disturbances affecting even the largest social groups. This is a growing problem, and while there is abundant historical precedent for the anxiety of our own period, one may doubt whether group panic has ever before carried so great a threat to the survival of society. Mass hysteria was common enough during the Middle Ages, as witness the Children's Crusades, the epidemics of St. Vitus's dance and other collective furies. Modern civilization "does not defend man against mental contagion and reactions of panic." On the contrary, advances in travel and communication, among others, have increased man's sensitivity, multiplied the carriers of mental contagion and augmented their effectiveness, and reduced the likelihood that a mental epidemic would exhaust itself in one locality before it had the chance to spread to others. Any knave or half-wit with access to a microphone can, in a few seconds, stampede half the world. It has been done and it will be done again.

"Knowing panic is, by and large, conquering and preventing panic." The forms and the causes of social panic are worth reviewing, even briefly, to help with our perspective of where we are and which way we may be going. Wild flight, rioting, vandalism are the most familiar manifestations; there are endless examples: the great Indian Mutiny, the Brooklyn Bridge disaster, the capsizing of the *General Slocum*, the massacre at Caporetto during the First World War, the collapse of the French Army in 1940. Panic also has its silent forms: in an overcrowded London air-raid shelter, in the spring of 1943, the lights went out as the result of a bomb explosion nearby. "There was a sudden upheaval of tremendous fear in the pitch dark; no yelling or crying was heard. When first aid arrived

175

nearly 200 of the 600 people were dead. Post-mortems revealed no significant anatomical changes in the victims." They had, it is supposed, literally died of fright.

Panic also has its ludicrous, pathetic and tragic aspects. Since itching has an "especially disorganizing influence on the human psyche," DDT helped morale considerably in the last war. In Quito, Ecuador, where Orson Welles's invasion-of-Mars radio fantasy was rebroadcast, the New York-New Jersey panic was repeated, except that fleeing street mobs were so enraged on learning of the hoax that they wrecked Ecuador's principal newspaper building. "On July 14, 1943, a British pilot bailed out over Holland. A Netherlander on a motorcycle was in the vicinity of the parachute landing and offered the pilot his help. The pilot, however, said that his instructions were not to endanger the Netherlanders but that he would like to be conveyed by motor to the nearest German post. The Hollander took him there. The British officer entered, followed by his escort, and the Germans were so frightened that they surrendered. When they discovered their mistake they started to scold vehemently." Inmates of the German extermination camps immersed themselves in and clung to "petty daily routines" to avoid thinking of the final horror. Whenever a group of Jews was sent to the furnaces, the same kind of scenes took place. "In the midst of panic and hysterical crying [I quote from the Dutch author De Graaff] there were people who remained eating their breakfast. Another explained, 'I cannot go now, my laundry is not dry. I have to take care of it.' A nurse stuck to her task and read all the thermometers over and over. A tailor remained sewing the coat of a German officer. They all wanted to deceive Fate by concentrating on little duties."

There is no simple relation, in the psychopathology of panic, between cause and effect. Panic is as irrational in its dynamics as in its cause, or at least must appear so, in the majority of cases, to all but the most skilled and sensitive observers. The Indian Mutiny started on a rumor, wholly false, that the cartridge cases, which the troops had to tear with their teeth so as to release the charge of powder for the muzzle loaders, were coated with either pig or cow grease: the first, a defilement of the Moslem troops; the second, of the Hindus. A melange of rumors, both spontaneous and carefully sowed by the enemy, contributed more heavily than dive bombers and blitz tactics to the collapse of the French Army in 1940. At Caporetto, when a huge Italian army fell apart in the face of much smaller enemy forces, it was ostensibly because the Italians had no gas masks. (The Austrians, to be sure, had no gas.) A feeling of betrayal or abandonment by its leaders may lead a whole nation to panic (the Dutch, for example, when the Queen and other government officials fled after the Nazi invasion in 1940), and almost the entire world was stricken with anxiety and depression after the death of President Roosevelt.

Economic and financial panics may be precipitated by sudden fears occasioned by the collapse of a bank, investment house, mining company or the like; their incubation period, however, is usually extensive, the embryo deriving nourishment from political or social unrest. Our economic relations, as Meerloo reminds us, are largely based on "mutual distrust and tremendous competition." An economic engine that must continually accelerate ("ever increasing prosperity") so as not to stall makes demands that few human nervous systems are able indefinitely to support. "On one side there is a mood of immense expectancy, of waiting for the golden eggs of the economic system, with its old frontier romanticism, its dream of wealth, but at the same time there is that hidden mental uneasiness about keeping up the tension, until the sudden breach in self-confidence takes place. Unlimited competition stimulates inner fears. People become suspicious because they project their own hesitations onto all kinds of social manifestations. After a boom the country may become more suspicious."

177

Meerloo points out also that *any* significant social or economic change gives rise to insecurity and panic. A new pattern of living, the abandonment of accepted ways, however discredited, means an alteration of habit, a venture into the unfamiliar; for the majority of men and women this is a precarious step. In part, at least, this explains the effectiveness of name-calling against measures and advocates of reform.

The more serious panics — in respect of scope and duration — are, as I mentioned above, those longest in the making. A mob quick to violence is also likely to be quick to disband. But long-persisting irritants — biological, social, economic — may produce something akin to chronic disease, punctuated perhaps by acute flare-ups. Illiteracy, superstition, bigotry, magic religious attitudes also play a considerable part in increasing the receptiveness to panic and in maintaining the panic state.

A valuable section of Meerloo's study is devoted to an account of the causes, course and consequences of contemporary panic. One observes a deeply involved and often contradictory pattern the true nature of which is only dimly apprehended; it may be that many of our fears are justified, that certain critical elements of the pattern have not emerged or are under control, that we have reached the crest of national anxiety and that the excesses that have occurred will act as purgatives and restore health. Nonetheless it is clear that the situation is dangerous, for, even adopting the optimistic view that we have passed the crisis and are convalescing, there is the ever-present possibility of a setback.

The American people have long been exposed to social and political conditions making for insecurity. The great depression; two great wars; the deep, unhealed social scars left by the incomplete surgery of the New Deal; economic conflicts; sectional and racial tensions; the tug of war between isolationist and nonisolationist tendencies; artificially stimulated xenophobia and the fear of mysterious new weapons (rays, discs, bombs, germs), ours as well as the enemy's; a sense of "atomic" guilt for having used the weapon and for building it now: these are among the principal old and new causes of unrest.

No less striking and characteristic are the contradictions of high opinion and policy we are asked to reconcile. It is said that we are prosperous and spiraling upward to new levels of opportunity and abundance; it is also said that unemployment is rising and a new depression may be around the corner. We are militarily the greatest power on earth; also, our Air Force is inadequate, our Navy obsolescent, our atomic monopoly broken, our atomic secrets dissipated and stolen. Our foreign policy is the only course that promises success in the struggle against communism; also, China is lost to us, in France, Italy and Belgium there is deep unrest, Korea, Greece, Persia, the Philippines and Indo-China are in various phases of disintegration, and the Marshall Plan seems to be approaching its goal no more rapidly than did Tantalus. We are internally secure, the Communists are a feckless minority and the FBI has its eye on every spy; also, the government, the labor unions, the education system, the entertainment industry are seething with subversion, the Communists are not only a formidable menace, of themselves, but are abetted by fellow travelers, liberals, dupes, socialists and Fair Dealers. We are well protected by the oceans against Russian attack, and, since intercontinental rockets are not a reality,* we need not fear atomic attack; also, another Pearl Harbor is imminent, our civil defense is in a parlous state and we should at once start dispersing our cities. We are firm in our adherence to the Bill of Rights, to our traditional freedoms (including the principle that guilt is personal); also, to meet the present menace we must be prepared to recognize and to punish such offenses as "near-treason," disloyalty, heretical sympathies and sympathetic association. We are dedicated to peace and to the dignity of man; also, it may be necessary to wage catastrophic

---

* This was written in the Golden Days of 1950. J.R.N.

war, risking the destruction of man, to preserve that dignity.

I do not mean to imply that all the dangers warned of are unreal, that there are in fact no antithetical situations, and no honest dilemmas. I mean to emphasize that there is abundant confusion, partly justified, partly the result of irresponsibility or deliberate design, that confusion breeds insecurity, that insecurity long continued is the precursor of panic. We are not yet, of course, so unstrung that panic rules in its grossest forms. But there is ample evidence of widespread agitation, of apathy, of mutual suspicion, of crumbling morale, of a creeping barrage of anxiety and fear, leaving behind it the ruins of some of our most cherished rights.

The blame for this state of things is often laid on the McCarthys, the Mundts, the yellow press and like creatures. Secretary Acheson has recently denounced the attacks upon him and his policies as "vicious madness." This is true, but it is also too simple. There is never a shortage of oafs in any trade; second murderers are readily hired. But what Acheson apparently fails to realize is that the attacks upon him are dangerous not because they are launched, but because they are seriously attended; that they are listened to because unrest and anxiety have greatly increased the receptiveness to reckless charges; that certain policies, opinions and alarms propagated by himself and by others are in part responsible for the assaults of which he and others are the victims. The attempt to scare Americans out of their wits by the bogeyman and greatconspiracy principle of history having been quite successful, it is now hard to know how to give reassurance. "Total diplomacy" does not, I think, leap to mind as the ideal method. A government that repeatedly alerts its citizens·to the danger of being swallowed by a monster from without or destroyed by a Trojan horse from within, and that formulates its entire foreign policy and much of its domestic program in terms of security, military preparedness and the winning of a cold and, if need be, a hot war, is in effect staging a race between a national nervous breakdown and an overwhelming popular demand for war — any war that offers the welcome substitute of dread reality for dread suspense.

The logic of human psychology, the logic of panic in particular, makes it inevitable that widespread anxiety should lead to widespread guilt. It has become a new virtue to suspect our fellow citizen, to project on him all the evils that are printed in the modern slogans about spies and traitors. Feeling guilty, we hope to purge guilt, to regain self-confidence by persecuting and punishing others. Feeling guilty and afraid, normal things become strange and hostile; the known becomes the unknown; the capacity for judgment vanishes. Whatever, under such circumstances, gives substance to shadow affords relief. "We may be 'traitors' or 'fascists,' 'spies' or 'Marxists'; the very accusation makes us guilty of a hideous crime, of being a mirror of another man's hatred and evil thinking." Thus it comes about that behind every false rumor "we find the deepest collective wishes of the community." It is this complex situation which the Secretary of State sought so flatly to dismiss.

Today we are at the point where fear of ourselves, of impending war, of world revolution, of unknown changes and forces, has reached alarming proportions. It is hard to see in this Salem-like atmosphere how reason is to reassert itself, how self-confidence and mutual trust are to be re-established, how men are to learn to measure and meet their awful real problems with maturity and courage.

"The intellect, the judgment are in abeyance. Life is running turbid and full; and it is no marvel that reason, after vainly supposing that it ruled the world, should abdicate as gracefully as possible when the world is so obviously the sport of cruder powers — vested interests, tribal passions, stock sentiments, and chance majorities. Having no responsibility laid upon it, reason has become irresponsible." Thus, George Santayana in 1913 — an ominous year.

181

PHYSICS
AND POLITICS

D ISCRIMINATING READERS will be grateful to
Jacques Barzun and once more to Alfred
Knopf for rescuing from undeserved obscurity this small
treasure-house of social insight, political sagacity and learning. Walter Bagehot's *Physics and Politics** is among the
nineteenth-century writings that provided both the methods of
analytic approach and the systems of ideas to form the underpinning of contemporary social sciences.

It is a study of the evolution of political communities in
light of the biological and anthropological discoveries of
Darwin, Wallace, Tylor, Lubbock and others; it maps the
significance for human history (as Max Lerner has said) "of
such forces as custom and revolt, innovation and imitation,
conflict and discussion"; it carries forward its often irresistible
dialectic in a style of the utmost clarity marked by periods of

* Walter Bagehot, *Physics and Politics*, with an introduction by Jacques Barzun,
New York, 1948.

Walter Bagehot.

irony and high eloquence. It is, in short, a work informed by the best of Victorian faith and thought, and adorned with its best literary manners. That is not to say it is without peculiarities and shortcomings, as will appear.

By "physics" Bagehot meant the whole of natural science, man's understanding and control over matter and energy; by "politics" he meant the social sciences, man's understanding and control over the relations among men. Mainly he was concerned with two questions: How may the principles of natural selection and inheritance be applied to political society? What are the political prerequisites of power? Thus, the astonishingly timely title of the book is misleading, if from it one should infer that Bagehot examines the kind of political problems arising from technological advances symbolized in our day by U235. He was, to be sure, dealing, as Barzun says, with power politics — of which he had an uncanny compre-

183

hension — but one of the unfortunate gaps in his treatment of social change is the failure to consider, except by superficial reference, the social relations and function of science and technology. Perhaps that is asking too much: at any rate, he brings to his subject the growing resources of nineteenth-century science, and, what is equally important, the fighting creed of rationalism expressed in Huxley's stern admonition: ". . . be prepared to give up every preconceived notion, follow humbly wherever and to whatever abyss nature leads, or you shall learn nothing."

In a half-dozen brilliant essays Bagehot traces the transformation of society from the "preliminary age" when polity was unknown, to the "age of discussion" when polity, good and bad, came to grow on every tree. For a starting point he takes the period of "such men as neither knew nature, which is the clockwork of material civilization, nor possessed a polity which is a kind of clockwork to moral civilization." It was the age of Dryden's dream: "when wild in woods the noble savage ran." Men were free, but freedom then was a nasty, brutish affair that meant being at the mercy of all when of mercy there was none. Such notions as existed of law and polity were at best "uncertain, wavering, and unfit to be depended upon"; the bases of social order and morality were unformed; men could not imagine the meaning of a nation because they could not think of themselves as being *alike*, much less perceive the advantage of common action for a common purpose.

As if it were necessary first to retreat, to get a running start, so to speak, in order to ascend the "steep gradient" to civilization, these early creatures, although at the apex of the Linnaean chart, were less amenable to organization, more fierce in their individualism, than animals of a lower order, among which co-operation for a joint cause, impelled by instinct if not by reason, is not uncommon. So it must have been, Bagehot thinks, before the value of obedience and of a "comprehensive rule binding men together" was recognized.

184

Before freedom as modernly conceived could be attained, says Bagehot, men had to pass through a period of surrender of will, of submission, to an elite enforcing custom, privilege, tradition, superstition and ritual. It was a period Bagehot might well look upon with profound distaste, yet, like Hobbes, he saw in it the beginnings of all that is good and valuable in modern society.

Before distinctions of race could be swept aside, men had first to know pride of race; before they could regain the halcyon day spoken of in Genesis when the whole earth was "of one language and one speech," their language had first to be confounded and it had to be said that "whoever speaks two languages is a rascal"; before they could innovate and break the "cake of custom," they had to learn to imitate and to bake the cake of custom; before being inspired to revolt against addiction to pattern, they had to grow weary of the pattern they had made; before men would discard armed conflict as the means of settling disputes, conflict had to serve its uses as an agent of natural selection. These were the bitter, monotonous trials that prepared the age of discussion. And if, like the men of the Constituent Assembly in 1789, we look upon that long past as no more than a "blunder — a complex error to be got rid of as soon as might be" — we fail to see, as they failed to see, that "that error had made themselves." For liberty itself is descended of tyranny: "Later are the ages of freedom, first are the ages of servitude."

How did men pass from the first stage of civilization into the second stage — "out of the stage where permanence is most wanted into that where variability is most wanted"? (Note how often Bagehot draws upon the language of biology to explain the processes of social evolution.) The old proverb speaks only of the first step, but in the making of nations it is the second step that counts. "What is most evident is not the difficulty of getting a fixed law, but getting out of a fixed law. . . ." The "arrested civilizations" — of the Far East, say —

illustrate the point precisely: having taken the first step but not the second, they look, says Bagehot, "as if they had paused when there was no reason for pausing — when a mere observer from without would say they were likely not to pause."

"That which doth assign unto everything the kind, that which doth moderate the force and power, that which doth appoint the form and measure of working, the same we term a law." And it is from law, as so defined by Hooker, that nations derive the permanence on which their existence, in peace or war, depends. But there must be something more than a bond to tie men together, more than a yoke upon all men and all actions. Perhaps it would be more accurate to say there must be something less. For whatever binds a society must not constrict its variability and its adaptability to new circumstances. The social organism, like the biological organism, must have freedom to move and breathe and change. As there is need for men who will impose authority and uphold the existing order, so there is need for others who will question authority and strive restlessly for reform. Progress is "in the prophets"; while the force of legality must hold the nation together, it must not go so far as "to kill out all varieties and destroy nature's perpetual tendency to change"; even rebellion and discord have their uses.

In preserving this variability and in contributing to "nation-making," many contradictory factors play a part. Progress itself emerges only from the interplay of opposites and the very conditions of progress are those that tend to arrest it. It is "fixed law and usage" that makes a nation, yet the fixity of law and usage keeps it stationary. In the formative stage, nations need the advantages of isolation both as a protection against disruptive forces from without and to gain the time needed to establish the framework of rule and authority within. Yet without communication, traffic and commerce with other nations, no nation can long survive. War is needed because it "nourishes the 'preliminary' virtues." Yet endless

186

wars must lead inevitably to utter ruin, and thus defeat even the purposes they may serve as operators on behalf of natural selection.

At last then, with the "head of the sage" helping the "arm of the soldier," a few peoples arrive at the age of discussion. Along the journey through time nations, empires, entire civilizations have risen and disappeared, leaving behind only myths, relics and crumbled walls. The age of discussion is "the age of choice." Athens was the first of the free states in history where sovereignty was divided, where men might, at least within limits, discuss issues of principle and decide upon a course of action based upon the outcome of their discussions. The fate of Socrates merely proved that toleration, as the idea is now defined — or at least as it was defined in Bagehot's time — had not yet won full acceptance. But the dialogues of Plato and the works of other Greek philosophers express the character of men who were free to reflect and probe, to inquire about all things, to believe that which is sustained by reason and to reject that which reason rejects.

"In societies customarily tyrannical, uncongenial minds become first cowed, then melancholy, then out of health, and at last die." And there are, as we well know, even shorter ways with dissenters. But in the age of discussion, or at least in the ideal image of such an age, there is a safe haven for dissenters: for it is an age of intellectual as well as political liberty, of free thought, of "advancing sciences." The desire of man to better his condition (Macaulay's phrase), which may be considered the source of progress, finds in such an age the necessary and sufficient prerequisites for its development. Originality is encouraged, fixity contemned; the trust in status is supplanted by faith in the dreams of the young and the wise; rational discussion dissolves the "sacred charm of use and wont"; the need for bigotry wanes, and thus its power. "The progress of *man* requires the coöperation of *men*." And this is only possible in an age of discussion, an age of trade, traffic, communication, science, toleration, freedom and peace. Still, the very notion of co-operation was "produced by one of the strongest yokes . . . and the most terrible tyrannies ever known among men — the authority of 'customary law.' "

One of the best features of this study is Bagehot's recital of the social and political problems besetting communities in their incessant struggle for existence. He seems to believe, however, that many of these are problems of the past; that having progressed, man is not likely to retrogress; that having entered the age of discussion, the forward nations of the Western world need not apprehend the reappearance of dark forces peculiar to earlier ages. Though he had the specific misgivings of one who knows he has "truncated" a problem because it cannot be solved in whole, and those generally of a critic and philosopher, of an intelligence both practical and subtle, he was not altogether uncommitted to the nineteenth century's faith in the ultimate perfectibility of society through the agency of reason. This century, I venture to say, has not vindicated that faith.

Take, for example, Bagehot's picture of the savage: "His life is twisted into a thousand curious habits; his reason is darkened by a thousand strange prejudices; his feelings are frightened by a thousand cruel superstitions." Is modern man so cleansed of these "monstrous images," so free of delusions and hallucinations, so impervious to the influence of fanatics, demagogues and mountebanks, so untouched by dogmas of race and nationalism, so discerning of falsehood and propaganda, so cured of violence and barbarism, that he may regard the folkways of savages with condescension and indifference? " 'These people,' says Captain Palmer of the Fiji, 'are very conservative. A chief was one day going over a mountain path followed by a long string of his people, when he happened to stumble and fall; all the rest of the people immediately did the same except one man, who was set upon by the rest to know whether he considered himself better than the chief.' " Are we so emancipated, so independent in daily habits, in social and economic practices, in our political behavior, that the anecdote cannot be regarded as a timely fable? "The ultimate question," says Carlyle, "between every two human beings is, 'Can I kill thee, or canst thou kill me?'" Is this a question merely for primitive man? Do not whole continents yet press for an answer?

It would be improper to take Bagehot to task for his errors of prophecy except in so far as his analysis of the process of social evolution suffers from an uncritical assimilation of Darwinian principles and analogies into his interpretation of social phenomena. Altogether, his was a balanced and wonderfully open mind. He had an extraordinary grasp of the perplexities of politics and was well aware of the vast difference between public and private virtues, the absurdity (as Barzun says) of trying to personify groups, classes and nations "as if they were solid entities with one will and one mind," the thinness of man's crust of reasonableness, the fact that government by discussion has from its origins been a "plant of singular delicacy," of the elusiveness of cause and the instability of effect in the political affairs of mankind.

Where he fails most conspicuously is in his disregard of the effect of economic forces on social movement and in his underestimation, as Lerner has pointed out, of the "democratic masses as a creative political force." He had no sympathy, Woodrow Wilson said of him, "with the voiceless body of the people, with the 'mass of unknown men.'" Thus, typically, he thought the English the most advanced people politically because of their "dull, fixed" habits of acting all one way (and then only slowly), and the French retarded politically because "a great deal too clever [and volatile] to be free." "Stupidity" he considered an essential prerequisite of political freedom. *Physics and Politics*, it might be said finally, is a penetrating examination of the law of civilization but an imperfect appreciation of the law of decay.

189

WAR OR PEACE

W HERE there is an ear for reasoned argument, these books will do a service.* They are quite different books by quite different men, but they share a common concern: the continued existence of the human species. A sociologist examines the forces thrusting the world toward war; a chemist assesses the effects of radioactive fallout; a group of scientists and engineers surveys the feasibility of an inspection system for disarmament; a physician pleads for an end to the nuclear-arms race. The tone of each book is also quite different: one sounds like a trumpet, one is cautious and analytical, one is restrained but remorseless, one is eloquent but gentle. All, however, reflect a deep awareness of man's peril: the probable annihilation of the human race in the event of another war.

* C. Wright Mills, *The Causes of World War Three*, New York, 1958; Linus Pauling, *No More War!*, New York, 1958; *Inspection for Disarmament*, edited by Seymour Melman, New York, 1958; Albert Schweitzer, *Peace or Atomic War*, New York, 1958.

190

One may differ with Mills's analysis of the causes of the next war, but it cannot be denied that he is a man of courage and of conscience. He knows what he wants to say, knows how to say it, and says it. He has written an angry essay. It is none the worse for that; a torch must burn to give light. He confronts three main questions: Do men make history and, so, wars? Are we drifting blindly into the final catastrophe, or are certain decisions and policies thrusting us toward it? Assuming it lies within our power to avert war, what ought we to do?

Take the first question. It is often said that certain great events such as wars are inevitable. Men are trapped by circumstances; fate makes the decisions. This ancient idea, gradually transformed into the belief that the Deity has fixed a Great Plan that determines every sparrow's fall, is, among thinking men at least, obsolete. Nature is hard but not malicious; the stars are indifferent. Resistless forces, it may be admitted, shape the universe, but man can see and judge and know and create and decide for himself. His freedom is bounded, but within its small sphere it is sovereign. The "thinking reed," though the weakest, is yet the strongest thing in nature. There is, however, another conception of fate that is not obsolete. It is in fact, Mills says, indispensable for adequate reflection on human affairs. According to this conception, history is "the summary and unintended result of innumerable decisions of innumerable men." The men do not form an identifiable class; the decisions are not in themselves consequential enough for the results to have been foreseen. Like crowds of particles the decisions collide, coalesce and add up to the blind result — the historical event, which, as it were, is autonomous. This is the fate Karl Marx had in mind when he wrote in *The Eighteenth Brumaire of Louis Bonaparte*, "Men make their own history, but they do not make it just as they please."

Thus understood, fate is not, in Mills's words, a universal

191

constant. It depends upon social structure and especially upon the concentration of power. Fate is a name we give to power so diffused and fragmented that we cannot discern the mechanics of its use; it is a name like probability, which baptizes our ignorance. But in a society where power is visibly concentrated, history is not drift or fate. It is the sum of the vital decisions made by the groups of men that hold power. What they do — or fail to do — makes history; what others do is of little account: they are the "utensils" of history makers and the "objects" of history making.

Given this diagnosis, we shall with Mills take it for granted that a handful of "high and mighty" of the Soviet Union makes history; but what about the United States? Mills's familiar thesis, already set forth in *The Power Elite*, is here reaffirmed. In our country history is made, he says, by a small group of men: "the high military, the corporation executives, the political directorate." This is the top level. The middle level is composed of "a drifting set of stalemated forces," and at the bottom is a passive, increasingly powerless "mass society." The power elite makes the decisions even though the formal democratic machinery has been relegated to the middle groups. The "great public" votes and even elects, but to what purpose? Of the men it elects only a few help make decisions and then only when they are admitted to the elite.

Mills now warms to his argument. The power elite (in Russia as well as the United States) is possessed by the "military metaphysic" that the constant threat of violence and the maintenance of a "balance of fright" are the essentials to a condition of peace. Many of the elite also believe that a war economy is the indispensable underpinning of economic prosperity. They are committed therefore to the arms race, which may in itself be the immediate cause of World War III. Political apathy of "publics" and moral insensibility of "masses" in both communist and capitalist worlds allow the economic and military causes of war to operate. Leading intellectual, scien-

192

tific and religious circles are either confused and permissive, or join in the cold war. Few intellectuals put pressure on the elite to change its policies; fewer still set forth alternatives. In framing this indictment, Mills is careful to say that the thrust to war is not an elite "plot," either here or in Russia. The elite of each country has its "war parties" and its "peace parties"; and both elites have what Mills calls "crackpot realists." They are men so rigidly focused on the next step "that they become creatures of whatever the next step brings"; they are also men who cling rigidly to general principles and who "join a high-flying moral rhetoric with an opportunist crawling among a great scatter of unfocused fears and demands." Practical next steps and great, round, hortatory principles, but no program: this is the main content, in Mills's view, of today's struggles of "politics."

Steadily we move toward the abyss. What is to be done? Mills's appeal is addressed mainly to intellectuals. They must stop fighting the cold war. They must make contact with their opposite numbers "among those now officially defined as our enemy." ("With them, we ought to make our own separate peace.") They must help educate one another. They must also remedy the default of religion and help awaken the public conscience, for religion itself is "morally dead" in the United States, and ministers of God, who are responsible for the moral cultivation of conscience, "with moral nimbleness blunt conscience, covering it up with peace of mind." Scientists should honor publicly those, like the eighteen German physicists, who have made their declarations for peace and against working on the new weaponry. Scientists "should attempt to deepen the split among themselves and to debate it." They should denounce secrecy. They should refuse to become members of a "Science Machine" under military authority. They should refuse to make weapons and boycott all research projects directly or indirectly relevant to the military. These are among the steps that Mills says would begin the practice

193

of a professional code, and perhaps the creation of that code as a historical force. The scientist by adopting it would reject "fate"; for he would thereby declare his resolve to take at least his own fate into his own hands.

Mills's book is primarily a polemic and a sermon. It is, as intended, provoking as well as provocative. Patriotic groups will not clasp his thesis to their bosom, and many disinterested students of society and power will differ strongly with his analysis of the causes of war. No open-minded person, however, will mistake this for the work of a hack or pitchman for the official lines of either side. Mills is his own man.

Linus Pauling, though gentler, is also his own man. His book is a powerful statement of the case for nuclear disarmament.

That war is today an insane method of solving disputes is a truth so obvious that it is hard to prove. Men are apt to acknowledge it, as they acknowledge their mortality, and then go about their business. But the proposition that we all have to die someday is not the same as that we all have to die the same day. Until now it had always been assumed that though men were mortal, man would endure. This assumption, as Pauling shows, has become questionable.

Five chapters of his book discuss fallout. They give one of the clearest summaries of the problem I have seen. While there is considerable disagreement as to what constitute "safe" limits of radiation with respect both to somatic and genetic injury, no responsible scientist doubts that wide "margins of uncertainty" (these words are from the UN report on radiation) surround all estimates. Before the subject became a hot military and political potato, it was accepted that even small amounts of radiation produce mutations, and that almost all mutant genes are bad. But because quantitative estimates of population exposure and of effects are so wide-ranging, it is easy for those who for official reasons wish to minimize the danger to becloud and confuse the issues. It cannot be foretold

194

whose genes will be altered nor in what way, so let's all stop worrying and keep our fissionables dry.

The physicist Edward Teller, who has sought and gained wide publicity for his views, is Pauling's prime example of a public misinformer. Teller regards radiation from nuclear tests as a negligible threat to world health. World-wide fallout is, in his cogent phrase, "as dangerous as being an ounce overweight." So far as possible genetic damage is concerned, Teller has stated that "radiation in small doses need not necessarily be harmful — indeed may conceivably be helpful." This puts fallout in roughly the same category as Lydia Pinkham's remedy. His views on radiation are, of course, essential to his general political position. He is convinced that the bigger the bombs, the better the chance for peace. If the next war kills everyone, the world will again be peaceful, but this is not, I think, what Teller has in mind. He pins his hopes to "clean" bombs, which would leave survivors. Continued tests are needed, he says, to develop such bombs, and the tests themselves are harmless. Another reason he gives for not suspending tests is that atomic explosions can be concealed; we could never be sure, therefore, that an international agreement prohibiting tests would not be secretly violated by the Soviets. Because of his eminence as father of the H-bomb, Teller's opinions have carried considerable weight. Lately their prestige has waned, Pauling's arguments having contributed to this result.

The statement that fallout is as dangerous as being an ounce overweight Pauling characterizes as "ludicrous." It is based, he says, on a gross statistical blunder and represents a 1,500-fold underestimate of the hazard. One of Teller's points is that the people of Tibet, though exposed to much more intense cosmic radiation than people who live at lower altitudes, are as healthy as we are (or, at any rate, used to be). Pauling replies that the Tibetan exposure to increased amounts of cosmic radiation should produce an increase in the incidence

195

of seriously defective children from 2 per cent — the average United States figure — to 2.3 per cent. But since there are no medical statistics for Tibet, Teller's statement is simply out of the blue. (He advanced it in a magazine article in order, as he admitted in a television debate with Pauling, merely to "quiet excessive fears.") Another of Teller's assertions is that the radiation danger to the average person is ten times as great from luminous-dial wrist watches as from fallout. Pauling demolishes this statistic. Teller, it may be felt, is better at bombs than at arithmetic.

Pauling gives other examples of official misstatements. An Atomic Energy Commission official wrote an article in 1955, which stated that "the total fallout to date from all tests would have to be multiplied by a million to produce visible, deleterious effects except in areas close to the explosion itself." He was later asked by a chairman of a Congressional subcommittee if he remembered how much radioactivity had fallen on the city of Troy, New York, following a test held in Nevada a few days before. The official, according to Pauling, estimated the amount as "something under 0.10 roentgen" and finally settled on 0.01 roentgen. The physicist Ralph Lapp, who was present at the hearing, pointed out that a million times 0.01 roentgen is 10,000 roentgens. Whereupon one of the senators observed that this amount would have the "visible and deleterious" effect of killing everybody in the region.

A commissioner of the AEC* said in 1955 that "the fallout dosage rate as of January 1 of this year could be increased 15,000 times without hazard." This is not as cheery as the million-fold figure, but cheery enough. Pauling is less reassuring. He notes that ten hours after the detonation of a small fission bomb at the Nevada test site, the level of gamma radiation at St. George, Utah was 0.004 roentgen per hour, which is at the rate of 0.1 roentgen a day. Multiplied by 15,000, this

* Willard Libby.

gives an exposure of 1,500 roentgens — a dose that will produce death within a few days from radiation sickness.

Pauling's own estimates of the effects of fallout are somber. He predicts an annual death rate of 8,000 if tests were to continue at the 1952–1958 rate. Moreover, 15,000 seriously defective children will be born each year in the world, whose defects must be attributed to the tests (this number does not include embryonic and neonatal deaths and stillbirths). It is possible, he admits, that his estimates are ten times too large or too small. Perhaps, then, only 1,500 children will have to be sacrificed annually to the maintenance of peace; on the other hand, if the figures are too small, 150,000 children may be required. Even if no more tests are carried out, a lethal legacy of carbon already released into the atmosphere will be handed on for thousands of years. According to Pauling's calculations, based to a large extent on data provided by the AEC, the quantities of this long-lived isotope discharged in the period 1952 to 1958 will ultimately produce about 1 million seriously defective children and about 2 million embryonic and neonatal deaths, "and will cause many millions of people to suffer from minor hereditary defects." Our generation will be remembered.

When in April of 1958 Pauling first called attention to the menace of bomb-produced carbon 14, he was sharply criticized by three geneticists in a letter to *The New York Times* for making "erroneous" and "exaggerated" statements, which "only add to the public's confusion and do not contribute to the solution of the problem." Commissioner Libby of the AEC dismissed Pauling's warning by saying that the effect of carbon 14 from nuclear tests "is equivalent to an increase in altitude of a few inches." But the ridicule backfired when the AEC's Division of Biology and Medicine issued a document, "The Biological Hazard to Man of Carbon 14 from Nuclear Weapons," whose conclusions agree quite closely with Pauling's. This document was unaccompanied by a press release, and no

197

enterprising reporter nosed it out. Lapp brought it to public attention in a letter to *The New York Times*.

Like Mills, Pauling makes a general appeal for peace and for international agreements on the cessation of bomb tests and on disarmament. His words are a *cri de coeur*, sustained by moral authority and reason. But there is little evidence that such pleas, uttered by those who know most about the dangers involved in a nuclear war, have persuaded leaders of states of the necessity of making radical changes in their policies. The argument is often heard that those responsible for the safety of the state are in a hopeless dilemma, for even if they are sincere in wanting peace and favor disarmament, it is impossible to frame enforceable disarmament agreements. If no inspection system is workable, as Teller and others have claimed, what is the practical value of test bans and treaties to outlaw weapons?

There are at least two replies to this argument. One is that of the noted German physicist, the late Max von Laue. Suppose, he said, "I live in a big apartment house and burglars attack me; I am allowed to defend myself and, if need be, I may even shoot, but under no circumstances may I blow up the house. It is true that to do so would be an effective defense against burglars, but the resulting evil would be much greater than any I could suffer. But what if the burglars have explosives to destroy the whole house? Then I would leave them with the responsibility for the evil, and would not contribute anything to it."

The other reply is less noble, but it is no less to the point in countering the untruthful and misleading statements that have made men skeptical about the possibility of enforcing disarmament. It is given in the Melman book, a co-operative study made at Columbia University, which provides a searching examination of inspection techniques. One of the contributors, the Columbia physicist Jay Orear, effectively refutes the claim that atomic tests can be "bootlegged." He demonstrates that a comparatively small number of monitoring stations uniformly distributed throughout the Soviet Union and within the United States, using the combined techniques for picking up acoustic waves, seismic waves, electromagnetic radiation and radioactivity, would constitute an adequate inspection system for test suspension. This, together with a provision that UN inspectors be invited to all large chemical explosions, should make it possible, Orear says, to detect all nuclear tests unless they are of such ultra low yield as to be in the class of World War II blockbusters. Substantially this view was adopted by the United States and USSR at the Geneva Conference of Experts held in 1958.

Some twenty papers, together with an excellent general report by Melman, make up the Columbia report. They deal with inspection of several major classes of activities: the production of heavy weapons of a conventional type such as tanks, artillery and trucks; ship and submarine building; aircraft, missile and fissionable material manufacture; the preparation of biological warfare weapons. Anyone who has thought about the problem quickly realizes that inspection of heavy weapons and naval vessel manufacture is relatively easy. The sheer mass of metal that must be moved and processed, the size of the plants and yards, the large labor force required, all facilitate controls. One cannot hide a whale in the back yard. Inspectors with access to major factories can ensure against large-scale evasions: the Columbia group estimates that 5,000 to 10,000 inspectors are needed for 493 ordnance and accessories plants in the United States. This number seems high, especially in view of the multiple approach recommended for every inspection assignment; but at least it represents a feasible operation, which would not interfere with normal industrial activity. Monitoring the development of biological weapons presents greater difficulties. Because their dependability as mass killers is uncertain, biological weapons are unlikely to be the first choice of countries with large stores of atom bombs;

199

but small countries with limited laboratory and manufacturing facilities may, it is suggested, avail themselves of what is called the "poor man's atom bomb" — dispensers of virulent material. "Continuous and close attention" should therefore be given these weapons.

These, however, are essentially secondary matters. The main worry is inspection of the production of fissionables and missiles. In addition to Orear's analysis of clandestine bomb testing, the book contains reports on the possibility of using established radiation protection services in an inspection scheme embracing the manufacture of nuclear explosives, on the possible "theft" of fissionable materials, on problems of inspection of missile components, propellants and guidance systems, on the "amenability of the air-borne propulsion systems industry to production inspection."

The contributors weave a pretty tight net. It is hard to imagine how any militarily significant evasion of a disarmament agreement could slip unnoticed through the inspection system here envisaged. The analysis is based entirely on publicly available information; no attempt was made to gain access to "secrets." In Melman's opinion this self-imposed limitation has given the report a "conservative bias"; that is, "more access and more knowledge might have revealed more strategic control points for inspection" and thus made possible the elimination of many points now regarded as essential.

The basic assumption is that between 200 and 400 large missiles could be used "to devastate effectively" any one of the larger land areas of the earth. A useful inspection system must cope with possible efforts to produce these weapons in clandestine ways and also with the problem of hidden inventories of arms produced before the inspection system was instituted. In Congressional hearings some years ago J. R. Oppenheimer and other witnesses made the point that enough fissionable material to devastate a city the size of New York could be toted around in a violin case. The purpose of this testimony was to point up both the danger of atomic weapons and the desperate necessity for nations getting together to prohibit their manufacture and use. But this somewhat over-heated illustration had unfortunate results. Those who accepted it at face value became convinced that no inspection system would work, that disarmament was altogether unfeasible and that the only prudent course for the United States was to amass an enormous stockpile of weapons. The Columbia report presents a more rational perspective. The contributors do not minimize the dangers, nor do they claim that any inspection system, however searching, is without loopholes; but the report offers convincing evidence that a comprehensive scheme can probably be framed, which would make secret rearmament so extraordinarily difficult "as to be virtually impossible." An inspection system is, after all, an alarm system. It is intended to give timely warning not of minor infractions but of illegal activities on a substantial scale, for these alone carry the threat of a decisive surprise attack.

The multiple-inspection approach recommended by the Columbia group includes among others aerial reconnaissance (useful to detect the production rather than the existence of missiles); analysis and auditing of government budgets; monitoring stations; checks on the whereabouts and activities of engineers and scientists; surveillance of plants producing air-frames, chemicals and fissionables; inventory validation. The contributors have tried hard to foresee how ingenious men might contrive to fool the inspectors. This carries the analysis pretty far. When serious consideration is given, for example, to testing — in connection with inventory validation — the alteration of records and the age of papers and inks, the reader begins to feel as if Dick Tracy had taken over. But these are minor aberrations, more than counterbalanced by a knowledgeable and sensible treatment of the various aspects of a tricky business.

It is no small thing to question, as this book does, the validity of the ruling notion of deterrence. It is received opinion, not confined to Mills's "elite," that genocidal weapons offer a reasonable guarantee of peace because no nation would deliberately commit suicide. But neither history nor social psychology unequivocally supports this opinion. People do not vote on going to war, and children are never asked. Deterrents may not deter, because the deliberate judgment that is essential to the if-we-kill-them-they'll-kill-us-so-let's-not-kill-them sequence rarely comes into play. Small causes may have large effects; moreover, the dropping of even a single nuclear weapon is manifestly more provocative than slicing off Jenkins's ear or assassinating an archduke. An accident can set a catastrophic nuclear war in motion, and as nuclear weapons are increasingly available and dispersed in more hands, the probabilities of such an accident must necessarily increase. "One aberrant, psychotic person or person gone momentarily out of control," Melman writes, "could explode nuclear weapons at a random place, or over any populated area. A space satellite could be mistaken for a ballistic missile." And when many countries possess nuclear weapons, if a warhead were set off in one city it might be impossible to identify the aggressor and therefore even to threaten retaliation. Thus the major assumption of the mutual-deterrence strategy falls to ground.

The best possible system of inspection techniques has weaknesses. A "foolproof" inspection system is a politician's catchword. The Melman report recognizes this truth and suggests a design to compensate for the inevitable gaps in inspection. This design, called "inspection by the people," counts upon the help of plain citizens of every country to enforce international agreements. Secret rearmament requires the participation of a large number of persons; among them there are certain to be some who are not in sympathy with the evasion efforts and who might therefore be expected to report them to the international inspectorate. Such persons must be encouraged and protected. If channels of communication from the population to the inspecting organization are always kept open, news of clandestine violations — the use of certain machines, the production of materials, the operation of prohibited processes — will almost certainly trickle through. A constant appeal urging the theme that "the international agreement is mankind's shield against mutual extermination and that a violation of this agreement is thereby a crime against humanity" would, in Melman's view, evoke a co-operative response in every country and make untenable the position of any government, or group of officials, found guilty of breaking the law. The help of such brave and good-willed men would be a most powerful adjunct to inspection.

Can it not be said that this is the most important conclusion of the Columbia study? In a sense it links together the different approaches of Mills, Pauling, Schweitzer and others who warn and strive to educate the world before it is too late. For after assessing the causes of war, analyzing the various strategies, designing meticulous disarmament and inspection schemes, one faces the irreducible truth that we can live together or die together. It is too much to expect men all at once to throw away their weapons and embrace. But a beginning must be made, and that beginning depends, as the engineers of the Melman report tell us, on conceding to each other what moral capacity we have, on having faith even in the enemy's awareness of his humanity.

"We cannot," says Schweitzer, "continue in this paralyzing mistrust. If we want to work our way out of the desperate situation in which we find ourselves, another spirit must enter into the people. It can only come if the awareness of its necessity suffices to give us strength to believe in its coming. We must presuppose the awareness of this need in all the peoples who have suffered along with us. We must approach them in the spirit that we are human beings, all of us, and that we feel ourselves fitted to feel with each other; to think and to will together in the same way. . . ."

ATOMIC HARVEST

W HEN a historian withholds important facts
likely to influence the judgment of his
readers, he commits a fraud, as Bagehot once remarked. When
a government withholds important facts likely to influence the
judgment of its people, it commits a graver fraud. Unfortunately these are offenses unknown to the law, though their
harmful effects are incalculable.

A prime example of official juggling — and of its consequences — is brought out in this book by Leonard Bertin,
science correspondent of the *London Daily Telegraph*.* The

* Leonard Bertin, *Atomic Harvest*, London, 1955.

204

subject of his account is Britain's atomic-energy industry. He describes its growth from small wartime beginnings to the great plants at Harwell, Capenhurst, Windscale and Calder Hall; the scientific and technological labors that made possible the expansion; the military and peaceful benefits that Britain hopes to derive from atomic energy. All this makes a readable and informative report; but the chief interest of the book for me lies in two early chapters, which present a timely recapitulation of a critical episode of recent history: the breakdown of the Anglo-American partnership in atomic energy.

Most of the facts, to be sure, are not new. They can be extracted from the Smyth Report, the official British and Canadian atomic energy statements of August 1945, Robert Sherwood's *Roosevelt and Hopkins*, Winston Churchill's war memoirs, and various papers and documents made public in the last few years. But the material is here for the first time pieced together, and while this is a partisan account, which exaggerates the British contribution in atomic energy, as others have exaggerated the United States contribution, it explains pretty well the nature, the causes and the results of the breach. It is a story worth pondering, more as a guide to the future than as another revelation, so fashionable in our period, of who killed Cock Robin.

To understand the culminating political fiasco it is necessary briefly to review the scientific background.

In January 1939, Otto Hahn and Fritz Strassmann of Berlin announced their discovery that an isotope of barium was produced by the neutron bombardment of uranium. A couple of weeks earlier Hahn had imparted this information in a letter to Otto Frisch, who was spending Christmas with his aunt Lise Meitner in Stockholm. Frisch, who was on the staff of the Institute of Theoretical Physics in Copenhagen, passed on the news to Niels Bohr, but not before making a profound and fateful conjecture. "It looked [Frisch told Bertin] as if the absorption of the neutron had disturbed the delicate balance between the forces of attraction and the forces of repulsion inside the nucleus. It was as if the nucleus had first become elongated and then developed a waist before dividing into two more or less equal parts in just the same way that a living cell divides." From an American biologist Frisch learned that the process by which bacteria and other organisms reproduce themselves is called "fission." He gave the same name to the newly discovered nuclear phenomenon.

Bohr brought the news with him to the United States and told his former student John A. Wheeler and others at Princeton about it. The word spread to Fermi and his associates at Columbia. Immediately several groups began a search for the massive pulses of ionization to be expected from the flying fragments of uranium. Within a few days the fission theory was experimentally confirmed in four laboratories in the United States, by Frisch in Copenhagen, by Frédéric Joliot (later Joliot-Curie) in Paris. The race was on: physicists in this country and in Europe drove themselves furiously to broaden their knowledge of the subject. Sensational stories began to appear in the press. When L. A. Turner of Princeton wrote an article on fission in the December 1939 issue of *Reviews of Modern Physics*, he cited and analyzed nearly a hundred papers on the subject published during the preceding twelve months.

The crucial question was whether a chain reaction was possible. This turned, as Henry D. Smyth pointed out in his account, on the result of a competition among four processes: escape of neutrons before capture, nonfission capture by uranium, nonfission capture by impurities, fission capture. One of the first groups off the mark were H. von Halban, Lew Kowarski, and Frédéric Joliot in France. They made experiments to ascertain the rate of production of neutrons, comparing the absorptive capacity of a uranium compound dissolved in water with that of a solution of a nonuranium salt. Their conclusions were basically correct but overoptimistic — for every neutron that splits a uranium atom, they calculated, between three and four additional ones are produced.

It was generally known by 1940 that to harness atomic energy, two different processes would have to be achieved, depending on whether one wanted power or an explosion. To get atomic power required a chain reaction induced by slow neutrons; to get a bomb required a chain reaction induced by fast neutrons, in separated uranium 235 or plutonium. It was also known that fission neutrons had high speeds, that the neutrons had to be slowed to produce power and to convert U 238 into plutonium, and that various substances — including graphite, heavy water and beryllium — could be used as moderators to slow down fast neutrons. On the basis of this knowledge the French scientists in 1940 made designs of heavy water reactors and graphite reactors to be used as sources of power. Lest anyone suppose that pure scientists are above sordid concerns, Bertin relates — not without relish — that the French team "hurriedly made the first application for patents of chain-reacting piles, not very different in general principle from those now operating in many parts of the world."

I need not devote time to the parallel United States developments during the period, since they have been often and fully described; but the British side of the story, much less widely known, merits some attention.

The importance of the Hahn-Strassmann announcement was, of course, not overlooked in Britain. Within a short time after the news from Germany, Sir George Thomson, son of the famous J. J. and himself a Nobel prize winner in physics, went to the Air Ministry and asked for a ton of uranium oxide. He explained that he needed it to conduct experiments on the possibilities of atomic power and atomic bombs at the Imperial

College of Science and Technology in London where he was teaching. The Air Ministry supplied the material and he began to work, but the first results were disappointing. He concluded that unless large supplies of heavy water were made available, a chain reaction in uranium oxide could not be realized. To the end of the war, fortunately (as Sir George pointed out to Bertin), the "most distinguished physicists in Germany thought the same."

But the most distinguished physicists *in* Germany were by no means the most distinguished physicists of Germany. Albert Einstein, Hans Bethe and Leo Szilard were in the United States, Lise Meitner was in Sweden, Otto Frisch was in Copenhagen, Sir Frederick Simon and R. E. Peierls were in Britain. These and other refugees — including Fermi from Italy — were to cost the master races dear. Like the Egyptians, they soon had cause to ask: "Why have we done this, that we have let Israel go from serving us?" In March, 1940, word came to Thomson from Peierls and Frisch (who had moved to Britain from Denmark) that a bomb made of uranium 235 was probably feasible. A committee was promptly set up under Thomson to go into the matter. The cover name selected for the group was *Maud*, by which choice hangs an amusing tale.

When the Germans overran Denmark, Bohr sent Frisch a telegram, the last part of which said, "Tell Cockcroft and Maud Ray Kent." Neither Cockcroft nor Frisch knew a "Maud Ray Kent" and concluded the words were an anagram for "Radium taken," intended to warn that the Germans had confiscated the Danish stocks of this valuable and significant commodity. At any rate it was the telegram that suggested the name for the committee. Years later, however, Thomson mentioned the fact to Bohr, who was astounded. It turned out that he knew a lady named Maud Ray who lived in Kent; he had wanted word sent to her that he was safe and well. Nor was this the whole of the comedy. For among civil servants who regarded themselves as being in the know, the name *Maud* was interpreted as a block of initials representing "Military Application of Uranium Detonation." These ingenious conjectures remind one of some of the hypotheses of cosmology.

Among the problems tackled by the Maud committee were the separation of light from heavy uranium, the physics of neutron capture and fission, the critical mass of U 235 needed to provide an explosion, and various related chemical matters. Peierls, Simon, James Chadwick, Norman Feather and the Swiss, Egon Bretscher, led these studies, which made promising headway.

Meanwhile a daring plan was conceived by that extraordinarily gallant and intrepid agent, the late Earl of Suffolk. As scientific attaché in Paris, he had compiled a list of 150 French scientists and technicians who were to be smuggled to Britain if France fell. When the Germans tore through the Maginot Line, the plan was put into execution, with Suffolk in charge. Halban, Kowarski and Joliot were on the list, but Joliot preferred to stay in France. In the confusion after the capture of Paris, only 40 of the men Suffolk had selected reached Bordeaux. There they were taken aboard a small British collier, the *Broompark*, for the hazardous voyage across the channel. Besides Halban, Kowarski and other scientists, the cargo included several million dollars worth of industrial diamonds, and thirty-six gallons of heavy water, comprising most of the then existing stock in the world. Suffolk, who numbered carpentry among his array of talents, built a raft, and the heavy water and diamonds were lashed on top. A "solemn agreement" was made that if the ship was mined or bombed, the survivors were to cut loose the raft and try to make port; but in the event of a torpedoing, in which case the Germans would have seen the raft, it was to remain tied to the *Broompark* when it sank. The *Broompark* arrived safely in Falmouth, although a neighboring ship exploded on a mine. The Earl of Suffolk,

209

afterward killed in attempting to disarm a German bomb, had brought off another brilliant coup.*

Upon landing in England, Halban and Kowarski, putting first things first, entered into preliminary negotiations to "safeguard their rights to any new discoveries." When this was settled, they set up shop in the Cavendish Laboratory at Cambridge and resumed their researches on reactors. According to Bertin they made notable progress, and by late 1940 had proved conclusively the feasibility of a chain-reacting, homogeneous, heavy-water reactor. If this claim is to be credited, they anticipated by more than eighteen months experimental demonstrations in the United States that a heterogeneous, graphite-moderated pile would sustain a chain reaction. The author says that when the Frenchmen's findings were "reported to the United States . . . they were immediately pooh-poohed." They are not mentioned in the Smyth Report, and it may be that even Chadwick's slightly qualified endorsement — he reported that while the claims were a little ebullient, their "general import was quite convincing" — did not suffice to silence our skepticism. Bertin seems to feel that nothing less than British and French honor were impugned by our attitude, and he emphasizes the fact — which is true, though the use he makes of it is malicious — that when the United States finally did catch up with Halban and Kowarski, it was the achievement of "foreign-born scientists working in America." This is petty

* Winston Churchill's account of the later exploits of the Earl of Suffolk is worth quoting. The Earl enlisted in the deadly game of trepanning and then deactivating time bombs. "One squad I remember which may be taken as symbolic of many others. It consisted of three people — the Earl of Suffolk, his lady private secretary, and his rather aged chauffeur. They called themselves 'the Holy Trinity.' Their prowess and continued existence got around among all who knew. Thirty-four unexploded bombs did they tackle with urbane and smiling efficiency. But the thirty-fifth claimed its forfeit. Up went the Earl of Suffolk in his Holy Trinity. But we may be sure that, as for Greatheart, 'all the trumpets sounded for them on the other side.'" Winston S. Churchill, *The Second World War*, Volume II, *Their Finest Hour*, pages 362-363. New York, 1949.

stuff, but it bred resentments that even now are apparently not forgotten.

The British work moved steadily forward, and there were frequent interchanges of information on their progress and ours. Members of the British Mission in Washington were invited to attend meetings of the National Defense Research Council subcommittee on uranium, and United States representatives in London attended meetings of Maud. By the summer of 1941 Maud made a high-level report on the possibility of a U 235 bomb and a pile to produce power and plutonium. (Three National Academy reports that appeared in 1941 showed the same trend of opinion, though the British were perhaps a little more confident.) Simon prepared a cost estimate for a full-scale gaseous diffusion plant; the figures frightened him and he decided he had better forward his lowest estimate lest the Cabinet itself suffer a chain reaction. After hearing the opinions of his advisers, Churchill drafted one of his classic minutes:

GEN. ISMAY FOR CHIEFS OF STAFF.

Although personally I am quite content with the existing explosive I feel we must not stand in the way of improvement and I therefore think that action should be taken in the sense proposed by Lord Cherwell and that the Cabinet Minister responsible should be Sir John Anderson. I shall be glad to know what the Chiefs of Staff think.

The Chiefs recommended a "maximum priority." A decision was taken to build a pilot plant for uranium in Britain and, if possible, a full-scale plant in Canada.

The Department of Scientific and Industrial Research set up a new unit called the Directorate of Tube Alloys. This cover name corresponded to our Development of Substitute Materials Project (in the "Manhattan District" of the Corps of Engineers). The group was first to have been called the Directorate of Tank Alloys — because Britain had become

211

"tank-conscious" as a result of the German blitz in France the
year before, but then Anderson hit upon "Tube Alloys" because a diffusion plant would require miles of tubing of a special alloy to resist corrosion. It was a nicely misleading yet
reasonable name.

It was one thing to create the administrative machinery,
quite another to erect the plant. Britain was heavily committed in manpower and resources, and the war was not going
well. Bombing attacks were frequent and the threat of invasion
was ever present. A United States proposal, therefore, to "co-
ordinate or even jointly conduct" the atomic energy program,
contained in a letter from President Roosevelt to Churchill,
was altogether welcome. The letter was sent October 11, 1941,
and it was followed up shortly by a visit from Harold Urey
and George Pegram to study the work going on in Britain. The
day before Pearl Harbor the U.S. Office of Scientific Research and Development announced to "S-1," its uranium
section, that a vigorous program would be undertaken to make
a bomb. British scientists were sent to the United States and
Canada for consultation, and Vannevar Bush advised a British
official, M. W. Perrin, that he favored the execution of a plan
to move the bulk of their research and development to Canadian laboratories. As a result, Halban led a team to Montreal. In June 1942, Roosevelt gave the go-ahead to our major
industrial project, and on August 13 the Manhattan District
was born.

It was generally assumed — or, at any rate, the British assumed — that the co-operation already achieved would be
extended, now that our gigantic program was going forward
and the major British effort had been transferred to Canada.
By autumn, however, it was clear that the channels of Anglo-
American communication on the subject of the atom, far from
growing wider, were becoming constricted. British scientists soon discovered that they could learn very little about our
progress on either the theoretical or the practical side. Fermi's

group in Chicago kept open a line to the Montreal group, but in general if the British were able to pick up anything at all, it was only out of the corner of some amiable fellow's mouth and not through official channels. In this respect, it should be pointed out, the British were not worse off than most United States scientists on the project, for they were strictly "compartmentalized." No one was told any more than high authority thought he needed for his particular job. This practice led to certain costly blunders but was regarded as essential to security.

In due course the breakdown of communications came to the attention of the Prime Minister. It was a painful surprise. From the beginning he had emphasized the necessity of maintaining a full interchange of information. He had discussed the matter with Roosevelt at Hyde Park in June 1942 and had come away with the understanding, as he afterward cabled Harry Hopkins, "that everything was on the basis of fully sharing the results as equal partners." "I have no record, but I shall be much surprised," he added, "if the President's recollection does not square with this."

At Casablanca, in January 1943 Churchill expressed his concern and Hopkins promised to look into the question on his return to Washington. This he failed to do, and on February 16 Churchill cabled, "I should be grateful for some news about this as at present the American War Department is asking us to keep them informed of our experiments while refusing altogether any information about theirs." Hopkins replied, asking that Anderson "send me a full memo by pouch of what he considers is the basis of the present misunderstanding, since I gather the impression that our people feel that no agreement has been breached." A long responding cable sent by Churchill, though he was sick at the time, recited the record of Anglo-American dealings on atomic energy and asserted that "fair play" demanded a restoration of the policy of joint effort. He concluded: "Urgent decisions about our programme

213

both here and in Canada depend on the extent to which full collaboration between us is restored and I must ask you to let me have a firm decision on United States policy in this matter very soon."

After further remonstrances by Churchill, Hopkins conferred with the President, Secretary of War Stimson, Bush and James B. Conant, who was Bush's *alter ego* in OSRD affairs. A memo by Bush (quoted in Sherwood's biography) stated that information would be furnished only to individuals "who need it and can use it now in furtherance of the war effort." To step beyond this policy, he said, "would be to furnish information on secret military matters to individuals who wish it either because of general interest or because of its application to non-war or post-war matters." This would "decrease security without advancing the war effort." On the surface this looked a sound position, but the British objection, as reported by Sherwood, was equally sound. Our policy, they said, gave us the excuse to withhold from them all the fruits of joint research, including the possible industrial uses of atomic energy after the war. In the event, this is exactly what happened. The security argument prevailed during the war; afterward, Congress, grossly misled by the Executive Department as to the nature of the British contribution and the Anglo-American agreements, passed legislation that forbade the transfer of atomic-energy data to any foreign country.

In the early summer of 1943, Stimson and Churchill discussed the atomic-energy impasse. Stimson took the position that the British pressure to obtain information was economically motivated and that "the Americans could not see the fun of spending billions of dollars to find out things for someone else to use in a competitive post-war world." At least this put the cards on the table. The British were deeply concerned, but further discussion was postponed until the heads of state met at Quebec a month later. There an agreement was reached, "typed out on Citadel notepaper," which provided, among others, for the setting up of a Combined Policy Committee "to keep all sections of the project under constant review" and for a "complete interchange of information and ideas on all sections of the project between members of the Policy Committee and their immediate technical advisers." The policy of compartmentalization was to be maintained in that exchanges at the working level were to take place only "between those in the two countries engaged in the same sections of the field." Bertin makes the point that this gave United States authorities the excuse to withhold information on production techniques since the British were not engaged in this section of the field. The facts seem at least partially to bear out this contention, for no British scientist or engineer, according to Bertin, "was ever allowed to enter the plant at Oak Ridge and our men were not even told whether the ideas that they had developed [on gaseous diffusion] had worked satisfactorily." Still, it is not altogether clear why, under the terms of the Quebec agreement, the British members of the Combined Policy Committee were denied access to these data — if, indeed, this was the case. (It is worth noting that while the United States representatives on the committee included Bush and Conant, the British representatives were Field Marshal Sir John Dill, and J. J. Llewellin, Minister Resident in Washington for Supply, who possessed no scientific training.)

But the most important provision adopted at Quebec related to the postwar uses of atomic energy. The fourth article stated that "in view of the heavy burden of production falling upon the United States as the result of a wise division of war effort, the British Government recognize that any post-war advantages of an industrial or commercial character shall be dealt with as between the United States and Great Britain *on terms to be specified by the President of the United States to the Prime Minister of Great Britain.* [My italics.] The Prime Minister expressly disclaims any interest in these industrial and commercial aspects beyond what may be considered by the President of the United

215

States to be fair and just and in harmony with the economic welfare of the world."

Whether or not this was a wise provision; whether or not it demonstrated — as Bertin contends — Churchill's magnanimity, or Roosevelt's shrewdness; whether or not it formed the basis for a workable agreement, which Congress would have approved, in the postwar period, one fact is certain: never was a solemn treaty between heads of state more completely disregarded than this one. The agreement was made in secret and for many years was kept secret. Churchill made it public in April 1954; Congressional leaders had been apprised of it somewhat earlier but not in time to inform their debates and decisions on the atomic-energy legislation adopted in 1946.

As a participant in some of these events, I am able to corroborate Bertin's statement that the McMahon Act was drafted, debated and passed by a Congress completely unaware of the terms of the agreement with the British and Canadians. While serving as counsel to the Senate Special Committee on Atomic Energy, I learned by chance, as did Senator McMahon, that there had been an agreement at Quebec. Repeated attempts to discover its contents were unsuccessful. The Senator and I called upon President Truman early in 1946 to see whether he could enlighten us. We came away empty-handed. Neither the White House nor the State Department thought it advisable to share their knowledge with a Senate committee. (We were not unaccustomed to this attitude. The War Department consistently refused to give the Committee essential information on atomic bombs.) Senator McMahon, according to Bertin, "to his great embarrassment, was only shown the text of the Quebec agreement after he had accepted the chairmanship of the Joint Congressional Committee on Atomic Energy" — i.e., in September 1946. David Lilienthal, it is stated, got the information when he was appointed chairman of the AEC. But this assertion is doubtful, since there is evidence that the members of the Acheson-Lilienthal committee, which drafted the famous control plan, had been told of the Quebec understanding.

It is unnecessary to read sinister motives into the decision of the Executive Branch not to make the legislature privy to the facts. An unblushing mendacity and astonishing high-handedness characterized this policy, but it is clear that disclosure would have loosed a storm on the administration; nor is it likely that Congress would have regarded itself as bound by President Roosevelt's more or less private and altogether vague commitment. Beyond question, Congress had the right to know of it, and the Executive had the duty to impart it. No one can say whether Congress would have acted differently if they had seen the agreement. Its provisions would not have been welcome, but at least our representatives would have been able to act with full knowledge, and the public would have been able to judge their action. Disclosure would also have countered the myth, still widely believed, that atomic energy was an exclusively American achievement and possession. Gordon Dean, in his book *Report on the Atom*, described the common misconception as follows: "Atomic energy was discovered and first developed in the United States in secret during World War II. Although we are still ahead in the field, the Russians, with the help of traitors, successfully stole enough of our key secrets during the war to develop a program of their own and are now hot on our heels. Our Allies, the British, because some of their scientists came over to help us with our war-time program, also know something of these matters, but are actually running a very poor third." No thoughtful person can fail to wonder to what extent the myth made more difficult the attainment of international control of atomic energy.

One asks why the British kept quiet about the Quebec agreement in the face of legislation "which made no distinction between countries like Britain and Canada and our late enemies or those with whom at the time we were engaged in a cold war." The answer is fairly obvious. The British desperately needed our help and co-operation in the years after the war. They were in no position to risk strengthening isolationist sentiment by ventilating a secret treaty. The British did what they had to do. They kept mum and built up their own establishment. And they did pretty well. Except in magnitude, their atomic-energy industry is at least the equal of ours — as to weapons, atomic power and isotopes.

Atomic Harvest is a valuable book. It contains errors of detail, improper emphases, and doubtful inferences. But the key historical account is essentially dependable and always honest. There is no reason to criticize its occasional flashes of anger. This is how the British feel and we may as well know it. Bertin is to be congratulated as a journalist who has done his job in the best traditions of journalism.

PART
4

THE
COMMENDABLEST
PHRASES

"P<small>ROVERBS</small>," observed John Hay, "are invaluable treasures to dunces with good mem-
ories." As one who can scarcely ever remember the proverb
suitable to the occasion, I am less cut by this unkindness than
perhaps I should be. For having just now finished reading the
second edition of the splendid *Oxford Dictionary of English
Proverbs** I must confess to being imbued with a great new
love of proverbs. Not that every proverb here presented is a
favorite; nor that every favorite proverb is here presented;
nor that one can fail to recognize that most of the proverbs

* *The Oxford Dictionary of English Proverbs*, compiled by William George
Smith, second edition revised by Sir Paul Harvey, with an Introduction by Janet
Heseltine, Oxford, 1948.

included are as dead as a herring, or, as one might say, as "dead as Queen Anne the day after she dy'd." Yet the book makes quite clear what was meant by Florio when he wrote that "Proverbs are the pith, the proprieties, the proofes, the purities, the elegancies, as the commonest so the commendablest phrases of a language"; and I not only join in Samuel Palme's lament but consider his the true explanation of the proverb's downfall: "Wise men make proverbs, but fools repeat 'em."

Janet Heseltine's introduction to the *Dictionary*, without being pedantic, is a learned and readable essay on the rise and fall of this form of literature and speech. An early example of the English proverb found in an eighth-century letter by Wynfrith, the Northumbrian missionary, is the familiar "Delays are dangerous," except that there it reads, in translation, "a coward [sluggard?] often misses glory in some high enterprise; therefore he dies alone."

The homilies and chronicles were full of proverbs taken from the Bible and from the writings of Church Fathers; soon they were enriched by materials of foreign origin extracted from Greek and Roman literature, from French and other Continental fables and romances. The art of rhetoric, so thoroughly cultivated in the Middle Ages, employed the proverb as one of its chief adornments. Proverbs were used to awaken attention in introducing a discourse or sermon; they served to point a moral in an age when men were not content merely to nibble daintily at didactic ethics. Medieval secretaries were well advised to keep on hand lists of proverbs (some of which have been preserved) for the drafting of letters on the formal lines "laid down by the *Ars Dictandi* of the schools."

By the middle of the fourteenth century, proverbs were being widely disseminated through the writings of Gower, Lydgate, Chaucer and their contemporaries. As a scholar who loved "to have at his beddes heed Twenty bookes clad in blak

222

or reed," and as an observant traveler who enjoyed being with people and carefully noted what they said and did, Chaucer was a master of the growing lexicon of platitudes. In the tale of Melibeus he could exercise, with modifications that reflected his genius as a poet and satirist, the medieval "love of sententiousness," which he shared, and thus he gave his hearers "a moral tale vertuous" with "more of proverbes, than ye han herd bifore."

Printed books gradually took over the role of manuscript or the spoken word; thus proverbs were ingrained even more deeply into literature. From Caxton's press came the *Dictes and Sayings of the Philosophers,* a dreary collection translated from the French, which nevertheless went through three editions in twelve years. Travelers on the Continent brought back Italian, Spanish and French works and, as Miss Heseltine points out, the proverbial wisdom of other peoples was put into "fresh English words" and absorbed, often without a trace of their origin remaining, into the writing and speaking habits of the English.

The stream reached its crest in Elizabethan times. The age was "soaked in proverbs." They were spoken, written and invented by scholars and wits, by the Queen herself and by ordinary men. John Heywood's *Dialogue Conteining Proverbes* and Thomas Wilson's *Arte of Rhetorique* were among the books studied by "every young writer of the day." Drayton wrote an unfortunate sonnet in proverbs; and in 1601 Thomas Jones, M.P., made a speech in Commons on a "Bill to avoid the Double Payment of Debts," composed entirely of proverbs. Though Jones favored passage of the bill, the speech is not to be recommended.

Yet of all these ardent practitioners, it was Shakespeare alone who used the old puns and the new, the dictes, sayings, maxims, aphorisms, adages "old say'd saws" and platitudes with distinction and originality. *Love's Labour's Lost*, Miss Hesel-

223

tine reminds us, was "crammed with puns, allusions, play upon words, and play upon proverbs, which would all at once be caught up by the audience of the day." In later plays he sometimes parodied prevailing literary fashions, but the proverb, more or less faithfully repeated, remains for him a remarkably effective and evocative instrument. "Constant dropping wears the stone" recurs in his plays in many forms; *Hamlet* has its share of old adages, and not merely those mouthed by Polonius; the jest made "by Lear's poor, shivering fool, when all around him he sees the elements, his master's wits, and his own world dissolving" ("And I'll go to bed at noon?") is the perfect example of Shakespeare's skill in the use of trite sayings: for no one but he, writes Miss Heseltine, would have "dared to let a heart break on a proverb."

Ben Jonson had a lesser opinion of the value of proverbs (they were useful chiefly, he thought, to "illuminate a cooper's or a constable's wit," and Mr. Downright in *Every Man in His Humour* has "not so much as a good phrase in his belly, but all old iron and rusty proverbs"), and indeed within a few decades this literary fashion of more than three centuries began to wane. By the time Chesterfield was advising his son that recourse to proverbs was proof of "having kept bad and low company," scholars and antiquarians had begun to collect proverbs and use was giving way to lore. The decline, to be sure, was neither sudden nor absolute. Old proverbs survived, new ones continued to be invented, and as recently as the nineteenth century they often were to be found in novels and in the writings of Victorian moralists. The literary conventions of our own period, however, as rigidly condemn the use of proverbs — except by way of irony — as earlier conventions required it.

At least three circumstances attendant upon this change of taste may be noted. The first, writes Miss Heseltine, is "that the history of the use and disuse of proverbs is a progression from the concrete to the abstract." The evolution of other literary forms, of all intellectual and speculative expression, for that matter, follows, with many turnings and regressions, the same direction. But in so far as the proverb is peculiarly a form of folk expression, it loses contact with its richest source as it departs from everyman's everyday experiences and observations. Though truly, how the cock, the sow, the rat, the louse, the weather, the wench, the lawyer, the dog, the doctor, the liar, the thief, the hypocrite, the horse, the priest and the goose behave is of no less concern to the scientist, philosopher and literary man than it is to the proverb maker.

The second point to remark is that proverbs, even if no longer uttered sententiously and with conscious purpose, have found their way into the idiom of the language and are used "insensitively." As the child is father to the man, so the proverb is the parent of contemporary modes of expression. It has played a not inconsiderable part in determining the rhythm, emphasis and timing of our literature and speech, just as it was formerly the surrogate of the ethical principles and precepts defining our culture.

And the third point is that for all the intellectual gains accruing to us by virtue of the trend from the concrete to the abstract, our spoken language is poorer in having lost the habit of the proverb. Only a bore or a senator will remark that Rome was not built in a day, that all is not gold that glitters, that not a stone should be left unturned, that misfortunes never come singly, or that a soft answer turneth away wrath. Yet there is reason to doubt that the manner in which these same sentiments are uttered today — and of course they are uttered, by logical positivists, by statesmen, bankers, newsboys and barbers — conveys the thought (if there be any) equally well. The coin of proverbs has been defaced by use and misuse; still the defaced token retains its metal and even the symbol of authority, where its modern counterpart of exchange is

225

often no more convincing than a mere bookkeeping transaction.

But now let me set down some examples, partially to illustrate what has been said. Observe how the proverb goes flat with refinement.

"An ass endures his burden but not more than his burden" is surely a better image than the insipid "Enough is enough." The banal aphorism about falling between two stools derives from the vigorous "Between two stoles, the ars goth to grwnd."

A favorite, which I was happy to find — "He will not change his old *Mumpsimus* for the new *Sumpsimus*" — is here fully glossed (see page 88 for clarification), and it occurred to me, as I saw it again, that it is the perfect rebuff for the salesman seeking to persuade you of the advantages of the new gyro-synchro-auto-hydro-turbo mesh. "Charity begins at home" is more vapid than Wyclif's "Charite schuld bigyne at hem-self," or even than "Charity and beating begins at home," from Beaumont and Fletcher.

In place of the infinitely dull articles on the problems of marriage, to be found as regular magazine fare, there is the succinct old adage: "When a couple are newly married the first month is honeymoon or smick-smack; the second is hither and thither; the third is thwick-thwack; the fourth the Devil take them that brought thee and I together."

Many proverbs that sound as though they had been recently coined turn out to be centuries old. "The Crocodile," said Lyly in his *Euphues* (1579), "shrowdeth great treason under most pitiful teares." It was Cato who warned, *"Inter os atque offam multa intervenire posse,"* more modernly: "Manye thynges fall betweene ye cuppe and the mouth," or "Much falls between the chalice and the chin," or "There is many a slip," etc. "I am ded as dorenail" was familiar in the thirteenth century; "A dogge hath a day" in the fifteenth; "Tell it not in Gath" was used by Wyclif ("Woleth ze not telle in Geth, ne telle ze in . . . Aschalon"); "In one ear and out the other" found a place in Chaucer's *Troilus* ("Oon ere it herde, att' other out it wente") ; getting a horse to drink when you had got him to water was a problem that had turned into a proverb by 1175; and "to kick the bucket" does not derive, say, from Chicago but rather from Norfolk (England) where the beam from which a slaughtered pig is suspended is called the "bucket" and thus for centuries the phrase has meant "to die."

The famous goose is older than the Borgias, the Piccolominis or the Percys; but of the many versions of her fate none is better than Lyly's: "A man . . . had a goose, which eurie daie laid him a golden egge, hee . . . kild his goose, thinking to haue a mine of golde in her bellie, and finding nothing but dung, . . . wisht his goose aliue."

"Who buies land buies war" is not altogether without its moral for us; nor is "All the weapons of war will not arm fear"; and "Who draweth his sword against his prince must throw away the scabbard" is a more eloquent statement of the problem of "disloyalty" than has fallen from the lips of J. Edgar Hoover. And to President Truman, concerned with the problem of subversion, I commend Bacon's "Lett Princes . . . not be without some person of great Militarye valew . . . for the repressing of seditions. . . . But lett such one, be an assured one, . . . orels the remedy is warse than the disease." "Auoyd your children," warned John Heywood in 1546, "small pitchers have wide eares" — still good by progressive-education standards and by the maxims of Freud.

In these pages I learned for the first time of "Pee and Kue"; of bringing "Pepper to Hindostan," "Salt to Dysart and puddings to Tranent"; of the advantage of inflicting upon an assailant a "Recumbentibus" (guaranteed "to smot in-two both helme and mayle") ; of coming to shoe the Pasha's horse and getting instead the beetle's leg; of Tweedledum and Tweedledee (originally used, it would seem, of two rival musicians) ;

227

of the "Ring of a Rush," the "Ring of Gyges" and the "Ring of Polycrates"; of time described as "a file that wears and makes no noise"; of not trusting "though he were my brother, He that winketh with one eye and looketh with the other"; of how and why "It would vex a dog to see a pudding creep"; of the Vicar of Bray and of the impossibility of doing two things at once ("A man cannot whistle and drink at the same time").

Also, it was high time to learn the origins of "No case: abuse the plaintiff's attorney" (c. 1600); or the early version "where greet fyr hath longe time endured . . . ther dwellth som vapour of warmness" (Chaucer); or that things were so hectic in the days of James I that it was thought "No newis is better than evill newis"; or that putting a man's nose out of "ioynte" was a figure of speech in 1581, while keeping it "hard to the grindstone" was a practice remarked on in 1532; or that "oisters" were not to be eaten, even in 1577, during the "foure hot moneths of the yeare, that is Maie, Iune, Iulie & August, which are void of the letter R"; or that it rained "dogs and polecats" in 1653; or that Lyly observed (?) the aversion of bulls to red rags; or that the Church was properly tagged for its venality in "No penny, no paternoster" (Tyndale, 1528); or that having "one foot in the grave" (i.e., "Thy graue is open, thy one fote in the pyt") was a common saying in 1509; or that "paying through the nose" meant paying through the nose long before the seventeenth century; or that the celebrated idiots' refrain, "In time of peace prepare for war," stems from Vegetius; or that long before the appearance of Dr. Fishbein's *Home Medical Adviser* (viz., 1500) it was commonly known that "The best physicians are Dr. Diet, Dr. Quiet, and Dr. Merryman" — obviously an even greater threat to the A.M.A., should the proverb leak out, than socialized medicine; or that Pliny, in Holland's translation, wrote, "When an house is readie to tumble downe, the mice goes out of it before," which Shakespeare in *The Tempest* gave as "A rotten carcass of a boat . . . the very rats Instinctively have quit it"; or that "Rich men are stewards for the poor" (1552) which is just, after all, what the N.A.M. always says they are, so why raise taxes?; or that "saving one's bacon" dates from the sixteenth century; or that Chaucer knew of "sixes and sevens" ("sexe and sevene"); or that the "twinkelynge of an yze" (Wyclif) was a measure of time even before Bulova; or that the "rap" was a counterfeit coin in eighteenth-century Ireland and thus not to give even a rap was to give very little indeed.

A fat, comfortable, impeccably edited volume of inexhaustible richness and variety. In gathering these samples of its contents I can only hope, as did Florio, that what was "a paine to me [will be] a gaine to thee."

A MILLION YEARS
FROM NOW

ABOUT twenty years ago William Olaf Stapledon, an imaginative English novelist, pub-
lished *Last and First Men*, a projected history of mankind
from the present to its presumed finish about five trillion years
from now. His narrative ended with the earth, inhabited by
the eighteenth species of *Homo sapiens* (reckoning ours as the
first), about to be destroyed by a solar convulsion. *Last and
First Men* is a book not easily forgotten. I returned to it when
reading Sir Charles Darwin's volume and found it as engrossing as on first acquaintance.

Compared to Stapledon's story the best science fiction today
is vulgar and paltry. Stapledon's book derived its power not
from its speculations as to the marvels of superscience, nor
even from its graphic descriptions of the cataclysms period-

ically engulfing the physical world, but rather from a remarkable treatment of the evolving nature of man himself. The subject is one that only the ablest writers venturing into this sphere of imagination, E. M. Forster and H. G. Wells among them, have been able to treat on a serious level. Stapledon was concerned primarily with the psychological traits of the successive human species. In his story one civilization after another grows up, flourishes and crumbles against a background of climatic changes, geological catastrophes, world epidemics, global wars, Martian invasions and the like. Again and again the human race is all but wiped out; then new species arise, often superior intellectually and physically to the races that have gone before. But the recurrent theme is the comedy of Eden: man is unable to let well enough alone; sooner or later he becomes bored and restless even in Paradise. Envy, curiosity, quarrelsomeness, suspicion, brutishness — sometimes dormant but never absent from his nature — reassert themselves and invariably encompass his downfall.

A similar theme runs through Sir Charles Darwin's *The Next Million Years.** Sir Charles, a distinguished physicist and grandson of the founder of the evolution theory, has not written a book of pure fancy and weird conjectures. His "time machine" is as dependable a vehicle as scientific knowledge can contrive. The million-year journey is conducted under the strict auspices of rational thought. The journey is no less interesting for this discipline. Reason and imagination are blended to provide a remarkably varied and stimulating itinerary.

It would be rash, as Sir Charles acknowledges, to predict the history of the next ten years. But every insurance company knows that long-range predictions are safer. Indeed, it is easier in some ways to say what will happen than what has happened. Assuming a moderate consistency in nature, it is possible on

---

* Charles Galton Darwin, *The Next Million Years*, London, 1952.

the basis of present knowledge to make certain fairly reliable forecasts over the next million years.

The classic example of large-scale prediction is Boyle's law of gases. We know almost nothing about the behavior of this or that gas molecule enclosed in a container before us, yet it is possible to predict the behavior of the whole ensemble of molecules with precision. Sir Charles proposes a Boyle's law of human behavior over the long run. To this end he avails himself of the same exquisite and paradoxical tool of prophecy, the theory of probability. To derive Boyle's law it is essential to know something of the internal conditions of the gas (viz., that the molecules constitute a conservative dynamical system, which is to say that the total energy of two colliding molecules is conserved) and of the external conditions, namely, the character of the containing vessel. The analogous law of human behavior, while not determined by statistical mechanics, also rests upon a knowledge of internal and external conditions. For the "wildly varying and violently colliding" gas molecules, Darwin substitutes the more complex human molecules, less varied, perhaps, in their trajectories but no less violent in their collisions. For the container, he takes the earth itself. On this foundation he builds a system of human thermodynamics.

The Next Million Years is not concerned with the trivia of history. The events historians consider important and stirring — wars, crusades, political upheavals, migrations, the crises of civilization, even such phenomena as ice ages — are passed over as of no consequence on this time scale. What matters is what will be happening "most of the time" on the endless train of summer afternoons. The aim of the book is to "form an estimate of the normal and not the exceptional course of the life of mankind on earth."

For the forecasting historian the crucial question is: who will survive? This must "override all questions as to whether future man will be better or worse than present man, or

232

whether he will rise to heights we cannot conceive or sink to levels we should despise." A species that dies out has by an inexorable principle demonstrated its unworthiness to live, however admirable or noble it may be by moral standards. "The dead shall not breed" is the dictum of natural selection; not the meek, but the survivors, shall inherit the earth.

In the long run, says the author, population must always press upon resources. Malthus's law is not mathematically exact and also must be modified by factors he could not have foreseen. Nonetheless it is clear that an unchecked population will outrun its food supplies. Undoubtedly food production can be increased substantially: tenfold, say, by bringing more land into cultivation and by improving agricultural methods; a thousandfold, perhaps, by ingenuities beyond present conception; possibly even a millionfold by intensive cultivation of the vegetation of the sea. Compare these estimates with reasonable expectations as to birth rates. It is plausible to suggest that in a century the world's population will have doubled. Wars are unlikely to arrest this expansion by much or for long. In three and a half centuries the earth's population will have increased by a factor of 10, in 10 centuries by a factor of 1,000, in 20 centuries by a factor of 1,000,000. Even with everyone munching on seaweed, Chlorella, sawdust and similar goodies, there will be barely enough food to keep the world's inhabitants crawling around. Yet 2,000 years is an insignificant period on the million-year scale. Even if the rate of population increase has been much overestimated, this hardly affects the argument, for assuming it took 1,000 years instead of 100 to double our numbers, we would still attain a millionfold increase in population within 20,000 years. We may take it as axiomatic that men will multiply, that food supply will not keep pace with their rate of increase and that a margin of the population must consistently be sloughed off by starvation, disease or other natural agents.

Sir Charles dismisses as unlikely any prospect that the hu-

233

man race will balance population and subsistence by deliberate limitation of its own reproduction. He sees no real chance that individuals will, in numbers sufficient to make a difference, practice birth control so as not to jeopardize the survival of their great-great-grandchildren. The processes that must win out on the long time-scale are those that are spontaneous and stable. The voluntary limitation of population, on the other hand, is a highly unstable process — the term "stability" roughly signifying that when a dynamic system "gets a little above its average level, by that very fact a force comes into play to pull it back, while if it falls below, a force is evoked to raise it again." Nations cannot be expected to accomplish much more by edict than individuals by voluntary restraint. To be effective, the limitation of the birth rate would have to be world-wide, based on a rational collective policy agreed to by all states. One need not comment on the unlikelihood of such a concordat. Though the majority of men might accept the policy of limitation on "broad rational grounds," one can safely anticipate a good deal of vehement, indeed fanatical, opposition based on creed. The main difficulty lies in enforcement, not only within the separate nations, but among them as well. A flagrant breach of the pact by a single country must disrupt a system that can maintain itself only by balancing its reciprocal pressures. Once the system collapses, only mass murder can restore the man-food equilibrium upon which survival depends.

We are led to a set of fairly simple and perhaps obvious conclusions, conclusions that the author's grandfather well understood. Men need food to survive; they must compete fiercely to get it. There are groups among them possessed of qualities that are likely to confer an advantage in this competition. These groups will determine the future course of history; their descendants will inhabit the earth a million years from now. It is hardly necessary to point out that the qualities destined to win out in the long run will not necessarily be those we now admire and seek to preserve.

What else can be predicted? The climate of the earth has been roughly the same for a billion years, and there is no reason to expect major changes for many more than a million years to come. A few ice ages might be disagreeable but could scarcely prove decisive. The planets revolve serenely, the sun continues to shine, and it is improbable that this harmony will be disturbed in the period considered. There is a chance, Sir Charles concedes, that a dark star may be moving toward our solar system. A collision would unquestionably do us in even if the earth itself were not hit, but the chance is so small that it need not concern us.

The problem of fuel resources is more serious. Our treasury of fossil fuels, accumulated for us over a period of 500 million years, has already been heavily depleted. At present consumption rates, it will be empty in 5 or 10 centuries. Future man must learn to live on income instead of on capital, and the standard of living is bound to fall. Water power will not fail, but it can contribute only a small part of world energy requirements. Atomic energy, from what is now known, offers no long-term solution. However, the potential supply of heavy hydrogen is in effect unlimited and if we could find a way to make it "burn" slowly, the problem would be permanently solved. There is scant reason to believe that ordinary hydrogen can ever be made to yield nuclear energy; if it could be, that, too, would yield a solution — of a kind. A frustrated *fuehrer*, or even a well-meaning lunatic bent on preserving the dignity of man, could set fire to the sea and to the earth's hydrogen envelope, with the result that the earth would for more than 10 years shine "as brightly as the sun does now." Such an event "would make the solar system into a very respectable new star," but plants and animals would be missing. Our descendants must look to other energy sources — sunlight, wind, tides,

235

vegetation, the interior heat of the earth, the cold water at the bottom of the sea — to eke out their needs. The expectation is not, in Sir Charles's opinion, altogether satisfactory, and mankind will probably have to learn to get on with a good deal less energy than our age is accustomed to.

Besides fuel, many other shortages will plague our successors. Stocks of essential metals will run out — some of them very soon. All sorts of ordinary things today regarded as indispensable, not merely to comfort but to existence, will vanish as the raw materials of which they are made gradually disappear. Substitutes will be found, but not for every object we have come to cherish; nor will the substitutes necessarily be of comparable quality. Plastic eating utensils are a devilish invention but supportable; the prospect of plastic surgical instruments and machine tools is disheartening.

The majority of the earth's present inhabitants may be justifiably skeptical of Sir Charles's contention that they are living in a golden age. Even the members of Western society may be permitted to doubt that this is the best of times. Yet in the material sense, disregarding certain inequalities of distribution, man is better off than ever before and better off than he is likely to be in the long-run future. Our period is marked alike by profligacy and invention; on balance we squander more than we discover, and thus over the centuries the estate diminishes.

I observed earlier that Sir Charles shares Stapledon's views as to man's essential qualities. In fact his book addresses itself primarily to the problem of "internal conditions": namely, whether man can be expected to change his nature markedly in the next million years. For if such changes were to occur, it would follow that several of the major inferences already set forth — notably as to population — would have to be reconsidered. "It was mainly the belief," Sir Charles declares, "that there will be no revolutionary change in human nature that emboldened me to write this essay."

He argues this proposition strongly. To begin with, he selects his span of prophecy, a million years, with certain geological and biological data in mind. These data force the conclusion that it takes a million years to make a new species. It matters little, apparently, how many generations occur in that period; some species change more quickly, some more slowly, but this "good rough rule" applies with general force to most of the species we know — rats, insects, buffaloes or men. For a million years, then, we shall have to get along with man as he is now.

It will not be denied, of course, that man has done much to improve himself since the species first appeared. How much more can he do? Sir Charles recites the four revolutions that have taken place in the development of humanity: the use of fire, the invention of agriculture, living in cities, and the scientific revolution. The "central fact" of this latest revolution, which we are still undergoing, "has been the discovery that nature can be controlled and conditions modified intentionally." The principal limits to such modifications — the constants and the bounds, in other words, of "external conditions" — have been discussed. It remains that the potentialities of science and technology stretch far beyond imagination; in this realm of conjecture the ground on which the prophet stands is certain to be treacherous. My feeling is that in these matters generally Sir Charles is too conservative. But I am delighted to report on his speculations on the use of high-speed computers as soothsayers. He sees the possibility of calculators that could improve on the performance of the Delphic oracle, not to say poll takers and military-intelligence experts. It may come about that such machines in a short space of time could "explore the consequences of alternative policies with a completeness that is far beyond anything that the human mind can aspire to achieve directly." Still, this brings us no closer to altering the drives impelling the man behind the machine. One suspects, for example, that a calculator which would reveal the outcome of a projected war would be discredited, if not banned, by sovereign states. So long as men believe what they want to believe, there will be wars.

Granted that no new species, "Homo sapienter," will arise spontaneously, does it lie within our own power to improve the existing breed of men as we have done with cattle and corn? Genetics affords no basis for an optimistic reply. The traits we regard as contributing to social good — tolerance, co-operativeness, rationalism — are "acquired characters" and do not pass through the sieve of inheritance. They are handed on by precept and example, not by genes; such, at least, is the authoritative consensus. It is possible to breed for milk-giving qualities in cows, for rust resistance in wheat, for learning capacity in dogs. It may be possible to breed men for pole-vaulting ability, for red beards or for aquiline features. But it is utterly beyond comprehension how permanently to build into the inheritance machinery the traits, say, of kindliness or healthy skepticism. Nor, for that matter, is it clear that such traits would necessarily promote the ability to survive. Changes can be brought about in the germ cells by means of X rays, which induce mutations. Startling effects have been achieved in this way with the long-suffering fruit fly, Drosophila. But this is no more than a hit-or-miss method that "simply stirs things up so that an arbitrary change results," usually deleterious. Tampering with the delicate mechanism of chromosomes and genes is like trying to adjust a fine watch by banging it on the floor; the treatment is unlikely to bring about the desired result.

The future, then, is fairly bleak. A million years from now the earth will be poorer, shabbier, more crowded than it is today. Its population, having been held in check by starvation, wars and other Malthusian governors, will be perhaps three to five times what it is now. The human species will not look very different. Its physical characteristics will change, of course, but "not to a great extent, since it is not primarily these qualities that preserve humanity in the struggle for life." We may assert with confidence that men will become "cleverer," for intelligence is a determinative factor in natural selection. Since Sir Charles does not share Socrates's view that wickedness is caused entirely by ignorance, he regards it as uncertain that men will become "morally better" as they become more intelligent. In a highly competitive world, moreover, the "sinner has many advantages over the saint," and whatever advantage promotes survival is its own justification and makes its own way.

Science, both pure and applied, will continue to advance. Inquisitive men will strive to know for the sake of knowing; others will build machines and invent practical processes. Sir Charles conceives of the invention of drugs to produce a permanent state of contentment (very useful for dictators), to remove the "urgency of sexual desire" and so bring about in humanity "the status of workers in a beehive," and to control the sex of offspring. None of these marvels, however, promises to add substantially to the sum of human happiness. Happiness, we are reminded, resides not in a state but in a change of state; eternal bliss is so tedious a prospect that one might prefer, given the choice, everlasting torment. Fortunately the dilemma is not thrust upon us. Life will continue to fluctuate between pleasure and pain.

The Next Million Years is a readable and honest book, packed with challenging ideas. The author speaks with authority on many matters, always with an attractive diffidence and in good temper. His explanations are clear, so clear that there is no difficulty in knowing where one disagrees with him and why. It is prudent to reject the notion that man is perfectible, but on the other hand it may not be overoptimistic to hope that he is capable of bettering himself before he vanishes from the earth. The Mendelian laws do not lock men into a groove of predestination; the science of inheritance, moreover, is not immutable. Man cannot be domesticated, says Sir Charles; he is a wild animal. In this wildness is his undoing, but also his glory.

239

MIND
UNDER MATTER

I T MAY BE that the experimental work of J. B.
Rhine in telepathy, clairvoyance, precognition
(prophecy) and psychokinesis (effect of mind on matter) has
produced the greatest scientific discovery of all times.* The
proposition seems doubtful if it rests on the case made in this
book; but if his results are valid, science faces a dilemma, not
to say a turn of irony, without parallel in the growth of organized knowledge. For Rhine's theories, if true, either de-
molish the conceptual framework of the physical world, or,
more alarmingly, demonstrate grave flaws in the mathematical
theory of probability, the cornerstone of modern scientific
method. The twist of irony arises in the form of a historical
coincidence.

* J. B. Rhine, *The Reach of the Mind*, New York, 1947.

240

The theory of probability has its origins in a seventeenth-century exchange of letters between Blaise Pascal and the upright parliamentary town councilor, Pierre de Fermat, on a subject assuredly remote from their way of life — an assessment of the correct gambling odds in games of dice. Professor Rhine, with equally lofty motives, now advances his views also on the strength of experiments performed with dice and cards.

Despite the respectable precedent for his methods, I find myself reluctant to follow Dr. Rhine into the beyond.

For almost two decades Rhine and his associates have been running their experiments on "ESP" (extrasensory perception) and "PK" (psychokinesis) at Duke University. During that period he has won disciples and been flattered by imitators, withstood jibes and criticism and defended his views in serious articles as well as semisensational books, of which this is the most recent.

His researches need not be described in detail, but they run something like this. Telepathy and clairvoyance, merged by Rhine into the single classification, extrasensory perception, are investigated with an absurdly simple tool, a deck of 25 cards, each bearing one of five symbols: star, rectangle, cross, circle or wavy lines.

In the telepathy test, after the cards have been shuffled, the experimenter picks up the first and concentrates on its symbol; the subject, from whom all the cards are well concealed, is then asked to identify it. The process is repeated until the pack has been exhausted, this constituting a "run."

The clairvoyance test is similar except that the experimenter does not look at each successive card until *after* the subject has identified its symbol. Presumably, therefore, the experimenter cannot concentrate on the symbols and "send" telepathic messages to help the subject.

It turns out that the errors one must guard against are almost incalculable. In the clairvoyance test, for example, if the experimenter himself is fortunate enough to be *both* a clairvoyant and a telepath (I trust this is the right word), he can detect the symbols by clairvoyance and transmit them by telepathy. For that matter, since space is no barrier to ESP, any friend of the subject, anywhere in the world, could help him out, assuming the subject requires help on a day when his telepathic reception is good, but his clairvoyancing bad. Rhine has anticipated such contingencies and by different controls overcomes them.

What are the results of the tests? Each symbol is repeated five times in the deck of cards, and the random chance of a correct identification by guessing is therefore one out of five. According to the mathematical theory, the ratio of actual hits to total trial should converge more and more closely to the theoretical probability (one to five) as the number of trials increases. If a subject consistently turns in an average of six or seven correct symbol-identifications per run through a large number of trials, the scientist should suspect and begin to look for factors *other than chance* that may account for the deviation. If a man won repeatedly at poker, one might examine the cards or his sleeve; a pair of perennially happy dice should be inspected to discover whether they are loaded.

Rhine contends that the over-all results of his ESP tests with cards show such significant deviations from the values required by theory that the odds lie from ten thousand to a million against one that the correct identifications are due to chance alone. There must therefore be other forces at work, unsuspected "powers of the mind," which, he concludes without more ado, are the fabled twins, telepathy and clairvoyance.

But these phenomenal aspects of ESP by no means exhaust its powers. The tally of the cards has also persuaded Rhine that ESP penetrates the future as easily as it traverses the ocean. His experiments have convinced him that some subjects are as adept at predicting the order of a pack of cards to be arranged (by random shuffling) *at some time in the future* as

242

others are in identifying symbols on cards already arranged. This function, which Rhine designates as "precognition," supports a pretty case for a naïve version of materialistic determinism, which, I had thought, had been quietly interred many decades ago.

Finally, the book describes Rhine's triumph, to wit: the discovery of PK. This function "of the whole mind" consists of influencing the behavior of matter by resolute concentration — what used to be called, in homelier times, "will power," except that Rhine has here measured its effect on the fall of dice instead of investigating its more usual range of reflexive operations, such as abstention from women, pastry, alcohol, shoplifting, tobacco and cognate evils. Among the resemblances between PK and will power detected by Rhine and his associates are those relating to the stimulating effect of rewards. Not only does a PK-er work much better when you give him a reward, but both ESP-ers and PK-ers turn into supermen when offered such inducements as "movie tickets" or "candy." On one lavish occasion, a young ESP-er named Lillian, transmogrified by a 50-cent bribe, "closed her eyes, talked to herself" and then crashed through with a *perfect clairvoyant score*, 25 hits in 25 cards.

One ultimate supersensation remains to be imparted, and Rhine is clearly the man to impart it. The final chapter, after roaming through Aldous Huxley, espionage via clairvoyance, man's misery and the atomic bomb, arrives at last at the topic of immortality, survival and the plausibility of "intercomcunication [of] discarnate personalities." This I take to mean both the small talk of the dead ("Do you miss us, Aunt Clara?") and that dismal repertory of spiritualistic japes which Charles Peirce ascribed somewhat unkindly to the "remarkable stupidity of ghosts." One learns with interest that Dr. Rhine has an open mind on the subject and may, if he wraps up ESP, investigate it. He will in that event find it nec-

243

essary to add slates to his present laboratory equipment of cards and dice.

If Rhine hitherto has been a victim of neglect, or worse, this book is hardly likely to relieve his suffering. He may be acquitted of the charges of being a fool, a faker, or both, but in these writings he plainly convicts himself of being a credulous purveyor of old wives' tales and superstitious nonsense.

Over the years, he has achieved a number of results that cannot be readily explained on the basis of conventional psychology. Certain statisticians and mathematicians have examined some of his experiment scores and found them, from their standpoint, unobjectionable. Now, this is no small achievement, and a patient, modest, persistent development of his methods might conceivably lead to important advances in psychology. But the achievement, and the applause, are too small for Rhine. It befits his temperament to make sensational generalizations and bizarre claims, in order to shock the scientific world to attention. He will not permit his experimental results to argue their own case; yet when he assumes the role of advocate, the effect is reminiscent of:

> But the judge said he never had summed up before
> So the snark undertook it instead
> And summed it so well that it came to far more
> Than the witnesses ever had said.

The jargon that Rhine and his school contribute to parapsychology endows this already doubtful subject with a suspicious flavor redolent of such disciplines as astrology, chiropractic and phrenology. My favorite example, taking precedence over psychokinesis, the psi capacity and the psychic shuffle, is "telepathic dermographia." This is a term describing an affliction, or, viewed differently, perhaps a talent, possessed by "Madame Kahl of Paris" who had (and for all I know, still has) the "ability to reproduce on her arm or breast, in clear red outline, a figure or letter of which the experimenter was thinking."

The jargon, of course, seems highly appropriate for someone who suggests, seriously, for scientific purposes, the use of such instruments as "automatic writing . . . the Ouija board, or the divining rod." Rhine's justification for the proposal is in itself a revealing gem: "If we are dealing with unconscious processes, why not use unconscious responses to record the key impression and keep the whole process unconscious?" Why not, indeed!

It would be selfish to take leave of this review without a few examples from Rhine's collection of clairvoyant and ectoplasmic chestnuts. No fable is too threadbare to be denied admittance to this hospitable book. In fact, Rhine is willing not only to suspend disbelief but to drop it altogether.

There is the story of the little child who, while out walking, is seized with a vision of her mother lying stricken and helpless on the floor at home, with a "lace-bordered handkerchief" alongside. In a frenzy of alarm the child rushes back only to find the vision confirmed, her mother *and the handkerchief* strewn about as expected. Commenting on this incident, which he extracted from a work named *Phantasms of the Living*, Rhine remarks, with commendable scientific objectivity, that "such non-experimental evidence" should be regarded as "more suggestive than conclusive."

I am also much intrigued by the case of the "world-famous psychiatrist" who, "just as he was about to begin to make a study of a medium located a few miles away," was greeted by a "pistol-like report . . . [which] consisted of the splitting of a very old, well-seasoned table top." Now, the psychiatrist was no fool and was not to be so easily deterred; as a scientist, after all, he knew that there are cases on record of wood splitting without a medium being in the neighborhood. But what really confused him was that an "old steel-bladed bread knife also went to pieces about the same time, accompanied by a

245

similar loud explosive report." "I was frankly puzzled," writes Rhine, but then adds cogently, "I have a photograph of the knife, which shows the blade clearly broken into four sections."

As I reflect on this book, I am harassed by a persistent, nontelepathic and nonclairvoyant vision of my own. During the war, among its many research projects, the War Department sponsored Sir Hubert Wilkins, the noted explorer, in a series of experiments involving the transmission of telepathic messages to a colleague situated somewhere in the frozen north. Sir Hubert would enter a little tent pitched in the middle of a large War Department office, taking with him some peanut-butter sandwiches, settle himself and proceed to concentrate.

Whether or not the results were successful I cannot say, but the performance made a great impression on me at the time and had at least one traumatic effect. I have since been unable to hear of telepathy, and was therefore unable to read *The Reach of the Mind*, without thinking of peanut butter. It seems to me only fair to offer this experience to Dr. Rhine.

QUEST
FOR THE MIND

T HE CLASSIC problem of the relationship between body and mind, the subject of endless
speculation in philosophy and extensive investigations in psychology, is here approached from the standpoint of its meas-
urable, electrical aspects.* The fact that the inquiry is scarcely
related in any immediate or obvious way to the more pressing
problems of our turbulent, paranoiac times does not diminish
its scientific importance. It would, I think, be better for the life
expectancy of us all if a somewhat larger portion of contemporary scientific effort were equally innocent in purpose.

The nature of mind and its relation to the operations of the

* E. D. Adrian, *The Physical Background of Perception*, New York, 1947.

247

brain are ancient and honorable objects of thought, but progress in mastering the problems has been insignificant. Almost insurmountable difficulties are encountered in the bare task of defining the basic concepts of mind and consciousness. Philosophers have manfully made the attempt, but their conclusions, even if esthetically persuasive, are of little use to the experimentalist. The use of thought to understand thought, the search by the imagination for its own essence, poses the highest challenge to the flexibility and resources of scientific method.

Our knowledge of the physical nervous system itself has more definite landmarks. Physiologists and anatomists have gathered detailed, accurate information about the brain, and especially about its outer layer, the cerebral cortex, from which hundreds of thousands of nerve fibers wind and twist through the body like the snakes of a delicate Medusa. Brilliant researches by the incomparable Sir Charles Sherrington, among others, have greatly increased our understanding of the integrative action of the nervous system, of the brain's functions as receptor and interpreter of messages from the outside world, as co-ordinator, to a large extent, of the reactions of the body itself.

Some of the more promising advances have resulted from the study of the electrical commerce of the nervous system. It has been established that the myriad nerve fibers in the human frame (and in that of other animals of higher order) are linked in a telegraphic system, over the channels of which messages are carried from any source of stimulus to junction points in appropriate regions of the cortex.

When the sense organs react to light, sound or touch, they initiate electrical waves that travel toward the cerebral cortex, one following another in series, their voltage constant regardless of the intensity of the stimulus, their frequency (i.e., the number of waves per unit of time) varying between 10 to 100 per second as long as the messages are being dispatched. Stim-

248

uli are distinguished as to source by the routes of their produced impulses and by the particular region of the cortex where they are received; they are interpreted as to content — much of this is still conjecture — by the frequency and total number of the electrical waves.

All nervous communication appears to be of this electrical character. In examining the behavior of the brain, the physiologist must in effect first "tap" its lines of communication. By studying the troughs and peaks of electrical activity in the cortex itself, he gains a one- or two-dimensional insight into what is probably a multidimensional phenomenon. He observes, for example, that the cortex produces *maps* — usually in miniature but sometimes, as in the case of certain optical phenomena, vastly magnified — of the events at the body surface. There is, in other words, an order and a fixed relationship between the points on the body and the points of the cerebral cortex — as phrenologists have so long insisted.

Consider how the mapping operation works in a simple case. When someone steps on my toe, my nervous system informs me of the nuisance by a signal. The pressure initiates a sequence of physical and chemical changes, accompanied by electrical currents culminating in a brain map. The map itself may resemble the flashing board of a railroad switching tower, giving the changing positions of the trains on the various lines. The electrical points of the brain map may actually form a tiny contour of the affected toe area. But our ignorance as to what occurs *after* the electrical map appears is profound. The brain certainly acts on the information conveyed by the map, but how? While the pressure is on my toe I may react in various ways, pursuing a combination of rational or irrational courses. The nature of the machinery and functions involved — whether I pull my foot away, upbraid the offender, suffer in silence or plot revenge — is obscure. The same is, of course, true in problems of a higher order arising in the trans-

249

mission and interpretation of (and reaction to) more complex messages, such as the contours of Lana Turner, the content of a page of differential equations, a Schönberg score or the image of a hockey match.

The most valuable projection area of the brain is where the maps corresponding to optical stimuli are formed. There stimuli may be magnified enormously, so that a small illuminated point produces an electrical map over a cortical area 10,000 times as large as the area of the stimulus.

Despite these advances in the study of the brain, little is known about the phenomenon of consciousness. At what point, if ever, do the electrical patterns of the brain rise into the region of *mind*? One series of discoveries, which may ultimately furnish a guide through the labyrinths of mental behavior, are those relating to the *cortical rhythm*. In 1929, Hans Berger discovered that electrical oscillations take place in the brain even when there is no apparent outer stimulus. If electrodes are attached to the skull of a nonstimulated, "inattentive" person, there is evidence of a steady 7-to-10-a-second rhythm of electrical wave frequencies (the α rhythm). Anything that provokes attention will change the pattern and the number of the oscillations.

Diagrams in Adrian's book show the flickering waves of the rhythm in a drowsy man, the waves suddenly becoming steep when the eyes are opened, and falling again as they close. The disruptive effect of auditory stimuli is also shown. There is reason to believe that the α rhythm is controlled by, or related to, the deeper brain centers where the "hobgoblin" who directs our consciousness has his perch. That this is a well-protected shelter is shown by the fact that even widespread injuries to the cerebral cortex, the outer sheath of the brain, do not necessarily produce loss of consciousness.

The electrical activities of the brain have been graphically and poetically described by Sherrington:

A scheme of lines and nodal points, gathered together at one end into a great raveled knot, the brain, and at the other trailing off to a sort of stalk, the spinal cord. Imagine activity in this shown by little points of light. Of these some stationary flash rhythmically, faster or slower.

Others are traveling points streaming in serial lines at various speeds. The rhythmic stationary lights lie at the nodes. The nodes are both goals whither converge, and junctions whence diverge, the lines of traveling lights. Suppose we choose the hour of deep sleep. Then only in some sparse and out-of-the-way places are nodes flashing and trains of light points running. The great knotted headpiece lies for the most part quite dark. . . .

Should we continue to watch the scheme we should observe after a time an impressive change which suddenly accrues. . . . The great topmost sheet of the mass, where hardly a light had twinkled or moved, becomes now a sparkling field of rhythmic flashing points with trains of traveling sparks hurrying hither and thither. It is as if the Milky Way entered upon some cosmic dance. Swiftly the head mass becomes an enchanted loom where millions of flashing shuttles weave a dissolving pattern, always a meaningful pattern though never an abiding one. The brain is waking and with it the mind is returning.

May we expect that further advances of science will permit of a purely mechanical explanation of the brain and of mind and consciousness? We have a reasonable knowledge of the cortex functioning as a screen, but the facts adduced do not confirm a naïve belief in the brain as a mere calculating machine. It is known that a tune may be recognized whether its pitch is high or low, that a triangle or letter may be recognized regardless of size or position. This is evidently something that can be "learned" as skilled movements can be learned. (For example, having learned to write with our fingers, we can then practice the art, even if not gracefully, with the toes.)

"These facts — of stimulus equivalence and of the transfer

251

of learned reactions — are a recognized stumbling block to all simple mechanical hypotheses of habit formation: they cannot be explained purely in terms of particular pathways which have been repeatedly exercised and have therefore acquired a special facility in transmission." It thus becomes impossible to represent the nervous system as a simple series of mapping machines unless an element be added that will discharge the higher function of abstraction. While machines may be constructed, at least in imagination, to react (by recognition) to a large number of set patterns, it is at present even theoretically impossible to design a machine that will go beyond a predetermined program. (If this could be done, I suppose our worst fears that the machine will one day take its place in the revolutionary scheme of things would have been realized.)

The brilliant Kenneth Craik, who before his early and tragic death worked in the Psychological Laboratory at Cambridge, made the ingenious proposal that thinking "depends on the operation of various nervous mechanisms which produce symbolic models of physical reality." Thus the models made in the brain would consist in part of the electrical maps of external reality, in part of symbols representing factors of experience and learning; figuratively, these elements could be shifted about like tiny pieces on a miniature chessboard, or like mathematical symbols during trial-and-error operations, for the purpose of envisaging probable future patterns of outside events and of formulating policies of action. The adaptability and plasticity of the models, the individual's ability both to make correct models and to preserve them in clear focus until their features and their implications were understood, would determine the clarity and effectiveness of his thought processes.

No one could be satisfied with this physical analogy for mental operations. Craik was certainly aware of its limitations. Yet it is an undeniable stimulus to further investigation along what may prove to be fruitful lines.

252

If ever the day comes when the operations of mind are clearly understood, we shall, I fear, still be plagued by the problem raised in Sir E. T. Whittaker's parable:

The observation of phenomena cannot tell us anything more than that the mathematical equations are correct: the same equations might equally represent the behavior of some other material system. For example, the vibrations of a membrane which has the shape of an ellipse can be calculated by means of a differential equation known as Mathieu's equation: but this same equation is also arrived at when we study the dynamics of a circus performer, who holds an assistant balanced on a pole while he himself stands on a spherical ball rolling on the ground.

If now we imagine an observer who discovers that the future course of a certain phenomenon can be predicted by Mathieu's equation, but who is unable for some reason to perceive the system which generated the phenomenon, then evidently he would be unable to tell whether the system in question is an elliptic membrane or a variety artist.

So far as what is happening in the world today is concerned, I have myself no confusion about the nature of the system. It is all due to a little man trying to keep his balance on a spherical ball. I even know his name.

THE WANDERING
ALBATROSS

"And I had done a hellish thing,
  And it would work 'em woe:
For all averr'd I had killed the bird
  That made the breeze to blow.
Ah wretch! said they, the bird to slay,
  That made the breeze to blow!"

AND NOW a mariner, neither graybeard nor ancient, has come to set things right. In November 1939 the pocket battleship *Graf Spee* had sunk a small merchant ship off Madagascar and then disappeared. Immediately the aircraft carrier *H. M. S. Ark Royal* was dispatched to find the raider — though where she had gone, east into the Indian Ocean or west into the trackless immensities of the South Atlantic, was not known. The Admiralty ordered the

254

Ark Royal "to take up a strategic position south of the Cape [of Good Hope] to cover either eventuality, but the seas around remained empty."

Among those who shared this lonely and weary vigil was a naval officer named William Jameson. For days, as the ship patrolled the horse latitudes, the sky was gray, the sea "lifeless and dull," and neither birds nor flying fish nor whales were to be seen. Then, on a fresh November afternoon, a few hours after the carrier had entered the sub-Antarctic zone, a great white form flashed into view. It was a wandering albatross, the largest of living birds. With motionless wings it skimmed the waves, turned into the wind, soared up to forty or fifty feet, banked toward the ship and swooped down at high speed to pass across its wake. For several days the albatross, which by now had picked up companions, followed the ship. On the edge of the roaring forties the sky became overcast, the glass fell and the weather turned wild. Wind velocity rose to sixty miles per hour; thirty-five-foot waves, three hundred to four hundred yards from crest to crest, made the huge ship lurch and plunge continuously. It was all a man on deck could do, in order not to be blasted overboard, to wedge himself into a sheltered corner. But the albatrosses were enjoying themselves: "swooping around at high speed, banking, soaring, diving to within an inch or two of the sea; perfectly at home in their boisterous element and moving, as it seemed, in any direction they wished." Then, as suddenly as they had appeared, the birds vanished.

For Jameson the brief encounter with these superbly graceful creatures was a "heart-lifting and unforgettable event." He continued his naval career, became attaché at the British Embassy in Washington, was promoted to rear admiral and knighted. When he retired he wrote a history of the *Ark Royal* (which after valiant service was destroyed by a U-boat in 1941) and then turned to a study of the wandering albatross. He searched the literature and assembled information from various other sources, and he has now written a delightful little book, which summarizes all that is known about this huge, gentle, fearless creature: its habits and life history, its method of flight, its place in fable and legend.*

The wandering albatross is of the family Diomedeidae, of the order Procellariiformes, which dates from the Miocene epoch (ten to thirty million years ago); fossil traces of an early albatross have been found in rocks dating from this period. Experts disagree on the number of different forms of albatross, but we may accept that there are some fifteen races. Three inhabit the northern hemisphere; the waved albatross is confined to the tropics. Nine species, which include the royal, the sky, the black-browed and the yellow-nosed albatross, have plumage that is mostly white at maturity, and have a range which overlaps that of the two races of wanderers. The wanderer, which is the largest species of albatross, breeds on bleak, stormy, remote islands: South Georgia, Inaccessible, Kerguelen, Gough, and others.

The name "albatross," says the *O. E. D.*, is apparently a modification of "alcatras," applied to the frigate bird, but extended through inaccurate knowledge to a still larger sea-fowl. (*Alcatruz* is the Portuguese word for the bucket of a water-raising wheel used for irrigation; whence *alcatras*, first applied to the pelican, which, the Arabs in Spain believed, used its great beak to carry water to its young in the desert.) The alcatras is black; perhaps to distinguish the white birds, the prefix albi-, albe-, (with etymological reference to Latin *albus* = white) was substituted. Sailors have called the wandering albatross by a variety of names: cape sheep (descriptive of its immense size and color), goney (obviously a reference to the very young birds, which have an innocent and befuddled look), man-of-war bird. In 1766 Linnaeus assigned to the wanderer a more poetic name: *Diomedea exulans*. According

---

* William Jameson, *The Wandering Albatross*, New York, 1959.

Albatross in flight.
(Photograph by Karl W. Kenyon, from National Audubon Society)

to Sir William, two fables guided the master classifier. An old sailor's legend says that sea captains are condemned in the hereafter to wander eternally above cold and stormy seas as retribution for the cozy quarters in the poop they enjoyed during their lifetime while the crew dwelt in wretchedness. The other fable concerns Diomedes, a Thracian hero, who after the Trojan Wars went to Italy, was ensnared by a magician and had his companions turned into birds resembling swans. *Diomedes* means Zeus-counseled; *exulans* means an exile or banished wanderer. Thus, *Diomedea exulans.*

Early books on natural history were neither very informative about *D. exulans* nor accurate. It was confused with other

257

birds, and strange habits were attributed to it — not stranger than the truth, merely wrong. In the late eighteenth century, as sailing ships sought out the "albatross latitudes" (40°-50° S) information about the birds began to come in. Mariners shipwrecked on breeding islands were more interested in eating albatross eggs — which are huge and delectable — than in compiling ornithological records, but now and then a more philosophical castaway controlled his appetite and enlarged scientific knowledge. In our century the expeditions of the *Discovery* and the *Quest*, and such leading ornithologists as Robert Cushman Murphy, Niall Rankin and L. E. Richdale have provided the data from which a good picture of the life cycle of the wandering albatross can be drawn.

On tiny desolate islands, scattered high in the southern latitudes, set in the wildest seas of the world, *Diomedea exulans* builds its nest. A favorite place is South Georgia where the birds, it is believed, have bred for thousands of years. Here, where it is bleak and treeless, where the wind always blows, where the foulest weather is the rule, where the only vegetation is coarse tussock grass; here, where huge sea-elephants and tiny petrels, seal and penguins come to breed and raise their young, the albatrosses gather for a few months to court and mate, to produce their young and raise them to independence.

The nest is built usually in an exposed place near the shore, facing into the wind, preferably on an eminence. The bitter cold and the constant gales do not discommode the albatross; even the nestling's feathers make a perfect coat. The main consideration is to choose a home that is also an airfield, where the birds can land and take off with the least trouble. In the air an albatross is grace itself, but on land it is a clumsy creature. Its toes point forward, it cannot perch, it walks laboriously with a waddling gait and has a hard time getting up enough speed to be airborne, and when it lands it is apt to pancake or to go over on its nose in a somersault. Nature is not kind.

Spaced twenty to fifty yards apart, the nests (made of soil, moss and grass trampled into a peatlike mixture) are shaped like miniature volcanoes. They are about three feet across at the base and are built high enough — one to three feet — to keep the top above the level of the winter snows. A single egg is laid in a shallow bowl hollowed out of the top of the conical structure. Husband and wife build the nest together — an arduous task — and keep it in repair; old nests are used again.

Egg-laying begins in November and December, depending on the island, and by early January (midsummer) eggs are everywhere plentiful. In 1923 four men from the *Quest* gathered 3,500 eggs in three days from the northern end of South Georgia; as an offset to such depredations the birds' food supply has been increased by the waste products of whaling stations, and the albatross population does not seem to have decreased. An albatross egg has a coarse white shell speckled with red spots, weighs around fifteen ounces and holds as much liquid as six hens' eggs. It takes roughly ten weeks for the chick to hatch, during which period male and female alternately sit on the nest. While one partner does brooding duty, the other is at sea getting food. The sitter is patient, fearless and tame; it will fight off piratical antarctic skuas, which are always hungry, but will permit a man to approach, provided he makes no sudden movements. *D. exulans* has a great bill that can inflict serious injury, but if this is firmly grasped, the bird can even be lifted from its nest and stroked. It is fair to conclude, says Jameson, that the wandering albatross is "devoid of aggressiveness." It only attacks what it supposes to be its usual food, but like a Yosemite bear it can be "dangerous without evil intent." (Park bears should not be fed because, as a keeper once told a lady visitor, they don't know where the bun stops and the arm begins.)

The newly hatched albatross weighs eleven to fourteen ounces and is covered with silky pure-white down. The beak (hooked at this stage) and the feet are yellow; the eyes very dark brown. In a week the chick almost doubles its weight, and in three weeks it is as big as a barnyard fowl. When it is a month old, it weighs more than six pounds and can frighten off any skua, so that it can do its own baby-sitting and leave both parents free to forage for it at sea. A new buffish-gray coat of luxuriant down keeps the youngster warm; autumn is now well advanced and there are frequent storms, but the bird is snug. More aggressive than its parents, it will snap at intruders; if this doesn't work, it will vomit its stomach contents, which are mixed with "a very evil smelling oil." This substance deserves a few words.

All birds of the tube-nose family manufacture this stomach oil, which is a gland secretion similar to the oil found in the head of the sperm whale. Its exact purpose is not known. It is, of course, a defense weapon, which can be discharged through the beak or nostrils; petrels are particularly adept in firing a stream of it fairly accurately and over some distance. It may also be a "supplemental food" for the young, since it is rich in vitamins. An intriguing suggestion is that ocean-going birds use it to calm the water in their immediate vicinity when riding out a gale. Jameson says this may sound absurd, but only to those who have not seen the pacifying effects of even a small quantity of oil on a turbulent sea. It is possible that the oil is used for preening, which is very important to creatures that spend so much time in the air. The oil contains alcohol, remains fluid at low temperatures and can easily be transferred from beak to feathers. A leading ornithologist has advanced the idea that the oil is the source of the fledgling's water supply. There may be truth in all theories.

At five months the youngster is still being fed. Although it is now fully grown, it is very fat, covered with down and cannot fly. Yet suddenly it is abandoned. Some time in June the

Another view of an albatross in flight.
(Photograph by Karl W. Kenyon, from National Audubon Society)

parents fly off to sea, apparently leaving the nest-bound fledgling to fend for itself. The ornithologist L. H. Matthews, visiting a snow-covered site in South Georgia in August, describes how each nest had "a young albatross the size of a small sheep sitting on it, sitting all alone and waiting; clad in a thick coat of buffish wooly down." A touching picture, almost inexplicable in biological terms, even if one recognizes that the birds can live a long time off their own fat, and that their metabolic processes during the rigorous antarctic winter probably slow down to a point closer to hibernation than can be found in any other creature of the bird world.

261

But this paradox of the life cycle seems at last to have been solved. By piecing together careful and arduous observations made in storms and in winter darkness at different sites on the breeding islands, ornithologists have now concluded that the fat young goneys are not entirely abandoned by their parents. After the initial phase of daily feeding, the interval is increased, at first to two days, then to longer periods. After the parents take to the wing in June, they return periodically to dispense fish and cuttlefish. This practice is continued until the young can fly. Two days before they go to sea, they get their last free meal.

The preparation for the youngsters' emancipation is spread over months. All through the winter the first coat of feathers has been growing underneath the second coat of down. The goney now looks rather frowzy, with tufts of down adhering to the feathers, but in due time its dress becomes more tidy. The sides of the neophyte's face are white, and there is a white bridge across the nose — "like a pair of spectacles." The feathers are a chocolate color; the beak is whitish yellow; the legs and feet, light-gray. A young albatross is "a somber bird," much different from the dazzling creature it will become.

In November it begins to flap its great wings — ten to twelve feet in span — but they are still weak. It waddles around the nest, makes trials standing into the wind, springs off the ground and lurches into the air for a few seconds. André Migot observed on Kerguelen that young birds took a month practicing to fly. Hubert Wilkins of the *Quest* party watched the birds painfully clamber up a hillside (resting frequently along the way), and then, "after a prolonged rest, run a few steps downhill and launch themselves into the air for a short flight, followed by a heavy landing which probably sent them tail over beak. Getting up, with an expression of pained surprise, they would repeat the process, always toiling farther up the hill if failing to get properly airborne." When the old birds begin returning in late November, the young exhibit their prowess and begin extending their range. For several months they make trial runs over the open sea. But when the breeding birds leave, the fully fledged youngsters set out on their own, and begin the "purely pelagic existence" that they will follow for several years. Some may not see land again until they return to their birthplace to mate and raise their own chicks; six or seven years may elapse before this happens. "Not for nothing has the bird been named the wandering albatross."

"It has always been a mystery to me . . . on what the albatross can subsist," wrote Darwin in his *Journal of a Naturalist*. The answer is that the wanderer has the huge pastures of the southern ocean on which to "graze." (This apposite descriptive term was used by Anders Sparrman, who accompanied Captain Cook on his voyage around the world in the *Resolution*.) It is a large bird and a surface feeder, but the area in which it can forage is immense: some thirty million square miles, or ten times the area of the United States. The ways of the wandering albatross are solitary. It has its own path, it feeds and flies and rests alone, many miles away from any other of its species. But when a ship appears, the birds are drawn together. In the days of sail the ruffling of the surface water, or refuse thrown overboard, attracted birds from every quarter; the modern ship leaves a propeller wake, and perhaps a trail of grease and oil, that acts like a powerful magnet. The *Tubinares* have a keen sense of smell, and their vision is such that they can see a ship eight to ten miles away. A dozen or more albatrosses may assemble and follow the ship for days. This brings us to the fascinating question: how do they fly?

Their aerial performance is prodigious. Consider: they fly into the wind at high speed; they move up and down in a curiously curved flight path, behind, over, alongside, in front of the ship, now trailing, now overtaking, now diving, now climbing, all the while scarcely flapping their wings; they can keep

263

up this pace for days, apparently without rest or sleep, following a vessel making twenty knots or more, and they have been known to cover a distance of at least thirty-five hundred miles in twelve days. Neither gravity nor fatigue seems to hamper these Zeus-counseled fowl.

The explanation rests on the fact that the albatross is primarily a glider or living sailplane. Though it is capable of powered flight, the bird's pectoral muscles are of moderate size and quite inadequate to support its sustained aerial feats. In a sailplane the force of gravity provides most of the motive power; the wings must give the required lift with the minimum drag, or resistance to forward motion. The aspect ratio (the ratio between the average width of the wing and its length) and the lift/drag ratio must both be high. In a high-performance sailplane the aspect ratio is around 18 — the same as that of the wandering albatross. The lift/drag ratio of such a wing may be as high as 40 to 1, "that is to say, for every forty pounds of lift acting upward at right angles to the wing there is a resistance to forward motion of one pound pulling to the rear."

To build a sailplane with such long and narrow wings, both light and strong, is exceptionally difficult, but the albatross meets these specifications admirably. Its wing bones are hollow and filled with air instead of marrow; in a twenty-pound bird the entire skeleton, including wings of twelve-foot span, weighs well under three pounds. The living creatures has other advantages over the man-made device. An albatross in flight can control its feathers and joints to change the span of the wings, their camber, sweepback and dihedral angle. The feathers can be closed or opened like a fan, thus enlarging or diminishing their surface, and achieving the optimum lift/drag ratio at any given speed.

A bird (or a sailplane) gliding through the air must overcome drag. This is done by dissipating potential energy. When the bird is high, it must slide "downhill" to pick up speed,

264

and somewhere along the descent it must find an updraft in the air stream to regain height. Currents of this kind occur over the land: on the windward side of hills, in "thermals" where warm air rises from the ground, and in "standing waves" where air passes over a hill, dips down and rebounds. All three types of updraft also occur over the ocean. Air currents rise on the flanks of waves, and rebound on the lee side of the crests. Thermals are found when the water warms cold air above it, causing the air to rise in closely packed counter-rotating cylinders; and even if the columns are blown over by the wind until they lie parallel to the surface, their counter-rotation pushes up a ridge of air along which a bird can glide.* The albatross seeks out such upcurrents and makes use of them. Yet together they are too irregular and operate too near the surface to afford the bird the steady and powerful support it needs.

Another source of power, however, is thought to be available. In 1924 P. Idrac of the Paris École Polytechnique carried out a careful field study, which led him to what Jameson describes as the now generally accepted theory of albatross flight. Wind blowing over the sea is slowed down by friction. In Jameson's happy image, if we throw a stack of playing cards onto a table they will spread out in different degree: the bottom cards, in contact with the table, will travel slowly over a short distance; the top cards fastest and farthest; the cards in between at intermediate speeds, each being held back by the friction of the slower-moving cards of the layer immediately underneath. So, roughly, with the winds blowing over water. For example, a wind traveling at 40 mph, 55 feet up, may be traveling at only 20 mph near the surface, and the layers in between will have a spread of intermediate speeds. The albatross capitalizes on this circumstance. It starts its

* This is well explained by John H. Storer in an article in the April 1952 issue of *Scientific American*.

dive toward the water, say at 30 mph from 55 feet, the wind as indicated. In still air it will accelerate to 48 mph, but since the air is moving and the bird is gliding downwind, by the time it reaches the slow-moving layers of air near the surface, it will have attained an air speed of 67 mph (i.e., 48 mph + 20 mph − 1 mph lost because of increased drag). This is more than enough to enable the albatross to soar back to its original height. Still, it would not enable the bird to follow a ship traveling into the wind, because it has lost considerable ground in its downwind dive. However, when it ascends and turns first across and then into the wind, it is constantly entering faster-moving layers of air, and the arithmetical balance of forces is in the albatross's favor, so that when it has climbed to 55 feet, it is actually moving much faster (perhaps 70 mph) than when it started to dive. The dynamics are a little tricky — and grasp of them is made harder by sundry leeways, leewards and windwards; but just as it is possible to sail into the wind, the albatross, making use of variations in windspeed within a shallow layer of air, as well as various eddies and updrafts close to a ship, executes a continuous, crudely S-shaped flight cycle, which enables it to keep up for days with a vessel making twenty knots or more against the wind. It is a feat no man-made sailplane can equal.

How long the albatross remains at sea after leaving the breeding site, how they navigate and find their way back to the tiny home islands, how often they breed: these are questions that are far from settled. Observers keep accumulating evidence but it is not decisive. The marvelous navigational ability of birds is of particular interest to the military in our happy age. On the rare occasion when they can blast one of their clumsy missiles off the ground and keep it flying, they would like to place it within a reasonable vicinity of the "target." What would they not give to have their wobbling, lethal toys emulate the swallow that flies from South Africa to the barn down the road, or *D. exulans* that circumnavigates

266

the globe half a dozen times and then homes to a nest on Kerguelen. Indeed, what treasure has already been spilled to gain this end. But if, as Jameson says, "the results of these investigations enable a rocket to home as certainly as the swallows of Capistrano, we must pray that the secret continues to elude mankind."

Jameson describes the remarkable rituals of courting and mating. The returning cocks form into groups of four or five, each composing a ring with a hen in the center. Hitherto-silent suitors now begin to utter "harsh, groaning sounds, not unlike the braying of an ass and no more pleasant or musical." The female faces each male in turn, and a dance ensues in which the cock shows what a splendid fellow he is. When the dance is concluded the female, whose deportment is demure, now turns to another bird and the ritual is repeated. Everything is carried off with decorum, "in a bucolic sort of way," and while there is a hideous din and much wing-flapping and neck-stretching, there is no fighting or interfering with the other chap's dance. Occasionally, when a cock's feelings become too much for him, he strolls a short distance from the ring with his head (as Robert Cushman Murphy describes it) "swaying from side to side and hung almost to the ground. The attitude gives [the cock] a diabolical look, and it would be easy to imagine that dark and sinister thoughts were occupying [his] mind." In time, after the female has inspected all the suitors, she makes a choice, whereupon the successful male shepherds his bride-to-be to his nest. "Other females are constantly arriving from the sea and the rejected males, undaunted, waddle off to try their luck elsewhere."

The ritual after pairing is even more elaborate than that of courtship. The birds croak at each other, opening their bills wide, and then nibble at the short feathers of their partners' necks. They emit "bubbling" noises, clapper, bring their beaks together and sing screeching duets. Standing on the nest, the male spreads his wings, stretches his neck, points his bill to the

267

sky, and turns slowly through a full circle, "lifting each great foot in turn high in the air with the measured precision of a guardsman doing a slow march. All the time he lets out a high-pitched pig-like squeak." The female, though "evidently greatly flattered," is still coy, but at last comes a demonstration that is irresistible. Stretching his wings to the full, cocking his tail upward, the male "shrieks to the full capacity of his lungs. No female could ignore so fervent a declaration." Mating may now take place, though quite frequently this "ecstatic climax is followed by anticlimax," both birds seeming to tire of the whole performance and sinking exhausted to the ground to rest quietly side by side. Then the whole ceremony has to be repeated, mixed in with a little nest-building so as finally to get the show on the road. The wandering albatross, "so solitary and aloof for most of its life, is an ardent bird, and courting continues, no less fervently, while the nest is being prepared and even after the egg is laid."

For the benefit of those who demand conjugal felicity even of birds before granting them their blessing, I should point out that there is evidence that wanderers are faithful to their chosen mates. But not in all cases and at best only up to a point. Promiscuity is known among the elders, and occurs even more among the youngsters. And each time an albatross returns to breed, it selects a new mate, the older males "showing the marked preference for sprightly young females which has sometimes been noted in other walks of life."

A final word as to the fable. Jameson has made a careful search of the origins of the fable upon which Coleridge presumably based his "Rime." The results are surprising. Birds have, of course, always been a subject of legend and superstitious awe. Eagles, vultures, swallows, geese, magpies, crows, ravens, owls have at various times and places figured as omens, as portents, as aids to augury. Gannets, kingfishers, petrels have had special meaning for seafarers, and many have heard of the famous cock that on April 12, 1782, perched itself on the

268

poop of Admiral Rodney's flagship and obligingly clapped his wings at every broadside poured into the French *Ville de Paris*. But of the albatross there is no record — neither of superstitions, omens, ceremonies, nor of cheering on the British Navy. If it was once believed that an albatross hovering about a ship brings bad weather, and that to kill it brings bad luck, the belief has left no trace. Moreover, there is ample evidence that sailors had not the slightest compunction about shooting an albatross, or snaring it by means of a tackle. The bird is no treat to the table, because it tastes like a stale, tough and oily fish, but catching it is sport — or what passes for sport. The albatross has also been prized for things that could be made out of it: tobacco pouches out of the webbed feet, pipestems from the hollow wing bones, a "handsome paper clip" out of the beak.

Where, then, did Coleridge get his notion? With the help of John Livingston Lowes's analysis in his famous *Road to Xanadu*, and some Coleridge correspondence, Jameson clears up the mystery — or comes as close to doing it as anyone would wish. I shall leave the reader in suspense; this is but one of several excellent reasons for turning to Sir William's charming natural history.

THE FORESEEABLE
FUTURE

H ow foreseeable is the future? Sir Charles
Darwin once took a long look ahead: he
leaped over no less than a million years.* Now another eminent
British physicist, Sir George Thomson, making a shorter jump,
peers into the middle of the next century.† On Sir Charles's
scale this is a modest prophecy, but it is not necessarily easier
to make. In ten thousand centuries the roundabouts make up
for the swings, averages assert themselves and entropy does its
dreary work. Darwin could count on the fact that even the best
machines run down. A million years from now, the earth, because of the voracious habits of its occupants, will be poorer
and shabbier. Its resources will be depleted, its oceans perhaps
desiccated. Men may be wiser but not happier; plankton is no
diet for joy. Maybe they will have to quit this planet for
another.

* *The Next Million Years*, by Charles Galton Darwin.
† Sir George Thomson, *The Foreseeable Future*, Cambridge, 1955.

Sir George's picture of 2050 is less doleful. He is chiefly concerned with the future of technology; in this direction, he says, predictions are possible. The scientific revolution has been churning for more than three centuries, during which enormous advances have been made in man's control over his environment. The rate of progress still seems to be accelerating. It is natural to ask how long this pace can continue. To be sure, a catastrophe may intervene, a man-made holocaust that would leave the future to rats or insects. We shall assume, however, that this will not happen: that either there will be no more big wars or the zeal for self-extermination will flag, once a war is on, before we are all dead. On this flighty assumption we can look forward to continued material improvement. We cannot say exactly where science and technology will carry us in a hundred years, but something of the future of machines is discernible in outline. The trend of social circumstance is more shadowy. "Sociology," says Thomson, "has still to find its Newton, let alone its Planck, and prediction is guesswork." Still, it is hard to resist the temptation to speculate on people's response to technical progress, and this book has something to say about the society of A.D. 2100 as well as the machines.

Science limits technology at the same time it feeds it. This fact is not widely understood. We expect too much of mechanical ingenuity. Technology has been so successful that men are apt to suppose that everything can be solved, answered, cured, that there are no limits. We have a light in the refrigerator, therefore we will fly to the moon; plastic seat covers promise plastic arteries; our theaters are air-conditioned, therefore we can change the world's climate. Let us grant that inventors are fertile and may make us immortal. But there are things they cannot do, things prohibited by the nature of nature as we understand it. So at least we are forced to think unless we are prepared to throw modern science out the window.

Sir Edmund Whittaker, I believe, coined the phrase "prin-

271

ciples of impotence" to describe the scientific rules of what
cannot be done. Since these are Thomson's guide lines, they
are worth listing. Perhaps the most familiar is Einstein's principle, that no material object and no signal can go faster than
the velocity of light. This should not unduly distress either
aeronauts or astronauts; nevertheless it sets bounds. The conservation of mass-energy is another limiting principle. Of
course the fact that this principle is a combination of two others, conservation of mass and conservation of energy, which
fifty years ago were thought to be distinct and immutable, is
disquieting and suggests that there are no permanent principles of impotence. But as I remarked a moment ago, we shall
have to do with the science we have, at least until another
Einstein appears. Sir George covers his tracks by saying that
while it would be rash to suppose the principles will remain
for all time, it would be "still more rash to suppose that they
can be modified in any particular way."

That one cannot make an electric charge or a magnetic
pole without making an equal one of opposite sign somewhere
else; that a particle of atomic size cannot be pinned down
exactly as to both velocity and position; that no two electrons can be close to each other both in position and in
velocity (Pauli's "exclusion principle," which keeps particles
of the same kind out of one another's way): these represent
three more principles to be added to the list. Finally, there is
the second law of thermodynamics, which says that "order
always tends to disappear till complete chaos is reached."

The foreseeable future of science, and of its offspring technology is thus bounded — though scarcely cribbed or confined
— by the principles of impotence. Moreover, it is reasonable
to suppose that other such principles may be discovered, valid
not only in physics and chemistry, but in biology also. As Sir
George points out: "Animals and plants have to be able to
reproduce and grow as individuals from a relatively very
small seed or egg, which yet contains the pattern of the whole.

272

There must be limitations introduced here. Not every arrangement of bones and nerves and muscles, even though it might make a visible animal, could, one would suppose, grow from an egg — still less be developed by evolution. Perhaps this is why nature never has produced a workable wheel or even a 'caterpillar' track."

Besides principles of impotence, there are other givens or fundamentals that must be taken into account. These define the scope of technology as a whole just as his materials and tools define what the individual craftsman can do. There is not an infinite number of different materials in nature, nor of forms, nor of building blocks. The marvelous profusion of the world is achieved with comparatively simple means. Less than a hundred different kinds of atoms (which themselves have only two or three common constituents), joined together in multitudes and in varied shapes, make the visible universe. A beech tree differs from a grain of sand, which in turn differs from a star; yet at bottom the three are alike and differ only as do mosaics. The fixed properties of the fundamental atomic particles — mass, charge and so on — impose limitations not only upon the particles themselves but also upon the larger bodies built up out of them. Man, for example, cannot grow beyond a certain size without altering shape. Beyond a critical point his bones will not support him; if he were as big as the moon, he would have to be spherical because "no material could make a neck capable of supporting such a head without being crushed." One thing is in man's favor in seeking to control nature. While it is true that he is "despicably weak" compared with elemental forces, the disposition of energy frequently enables him to take advantage of what are called "trigger" actions, "where a small cause produces a disproportionate effect." A boulder perched on a ledge may be pushed off easily, and produce an enormous avalanche. A handful of silver iodide may produce rainfall over a big area. There is a large class of such "metastable" systems, organic

273

and inorganic, requiring the turn of only a small key to unlock their energy. The right key is, of course, not always easy to find, nor are the consequences always easy to reckon. "A very high degree of understanding," Thomson warns, "is needed by those who would interfere with nature in this way."

Having set forth briefly some of the limitations on technology and told us what will not happen in the next century, Sir George undertakes to predict what can be expected. He discusses the future of energy and power, of materials, of transport and communications, of meteorology, of food, of applied biology, of social studies, and, finally, of mechanical devices for solving problems far more complex than any we are able to tackle at present. What he offers is a brilliantly succinct review of the main questions of contemporary science and technology, together with clues and conjectures as to how they will be answered.

Many books and articles have appeared in recent years on future needs and sources of power. Until nuclear energy was discovered, the outlook was bleak, Even now there are experts who are skeptical about atomic power. Sir George puts matters into perspective and resolves many doubts. He makes several things clear: first, that supplies of fossil fuels, coal in particular, are still very large; second, that we are so wasteful and incompetent in generating power that some of our processes might properly be regarded as no better than burning the house down to roast the pig; third, that the potentialities of solar energy and of combustible materials (such as peat) raised in agriculture require much more serious consideration than has been given them; fourth, and most important, that electricity derived from nuclear reactions, both fission and fusion, will be available in almost any quantities that we want for a very long time. This is not to say nuclear power will necessarily be cheap. Technical improvements will bring costs down, but large-scale projects will still be expensive in terms of capital investment, depreciation and the like. Efficiency will therefore be at a premium.

Shipping will "go nuclear" about the time that natural oil gives out. The internal combustion engine is already an anachronism and only waits upon the invention of a satisfactory electrical accumulator to become wholly obsolete. The heat pump, says Sir George, is a device of much promise, especially for the heating of homes. It is "really a refrigerator in which instead of creating cold inside the refrigerator by taking the heat away and then discarding the heat, you have the refrigerator out of doors and introduce the heat into the house." With cheap electricity as the power source, it is "the obvious way" of keeping houses warm.

In sum, whatever shortages may arise during the next century, power will not be among them. This is less obviously true of materials. Our civilization would be quite different if it were deprived of the new materials introduced in the last fifty years: light metals, plastics, steel alloys. Thus far we have been fortunate in finding great concentrations of valuable metals, but even the richest lodes are not inexhaustible and it will soon be necessary to search for and exploit the leaner deposits. This demands improved methods of prospecting, of mining at great depths where it is very hot, and of handling enormous masses of material in order to extract very little. Sir George suggests the possibility of deep mining, using the techniques of oil drilling. Under the crust of the earth lies molten magma, a pasty primordial mixture of minerals or organic matter. No one knows the extent of such veins of liquid, but if they could be located it is conceivable they could be tapped and brought to the surface.

A more promising source of minerals, perhaps, is the "fluid ore" composing the ocean. Concentrations are very low, but quantities are immense, and the fluid ore is easily handled. A remarkable feat is the ability of certain sea animals to extract from the water metals forming an essential constituent of their blood. Copper occurs in sea water only to the extent of two parts in ten thousand million, vanadium slightly more. Yet there are fish and other creatures which successfully concentrate these elements in their body fluids. It may be possible, Sir George believes, to breed plants or other organisms that will perform for us the first stage of extraction; then we can take over with our clumsier methods.

Much can be done to increase the strength of materials. Metals, and other materials, fail by shear, that is, by slipping of one layer over another. It had been supposed that if large single crystals could be made and tested, they would show great strength because the slipping characteristic of a large aggregate of small crystals would not occur. But when large crystals were finally made, it was discovered that they were extremely soft and could be sheared and deformed with very small forces. Research disclosed that the cause of this property was "dislocations," structural faults within the crystal. Hardening processes are believed to take advantage of these dislocations; that is, the working of the metal produces "dislocations in different directions which lock against each other" and so block slippage. But the dislocations are still there, and the jigsaw arrangement is far less efficient than an array of faultless crystals would be. There is no reason to believe, Thomson says, that these elements of weakness cannot ultimately be eliminated.

Once materials can be manufactured with much higher breaking stresses, a drastic change in structures will result. Some machines require thick and heavy members not so much to keep the machines from breaking as to keep them rigid. Increased strength of materials without a corresponding increase in stiffness will not help in such cases. But many other structures — buildings, airplane wings, cables of suspension bridges — can tolerate considerable flexibility as long as they don't break. Here the new materials, light and strong, should prove invaluable. The buildings of the twenty-first century "may be a little like the masts and rigging of a sailing ship, with the spaces between the structural members enclosed with a light 'cladding' of which a considerable fraction will be transparent." Sir George predicts that "the world of the future may be expected to look more aetherial, more like fairyland, than the world of the present or of the past."

Communication has come a long way in the last fifty years, and further improvements are foreseeable; many of us, however will be glad to hear that for various technical reasons it has its limits. The walkie-talkie will undoubtedly proliferate as transitors are improved; thus privacy will suffer additional indecent intrusions. Television is another device whose uses will be much extended. As an entertainment medium it can scarcely fail to improve, assuming any change, but Sir George is not concerned with this aspect. We can anticipate face-to-face "meetings" of groups of people, while each person remains in his own home or office. Business, political and even scientific gatherings could be arranged in this way. Yet one dares to hope that the need and desire for actual human contacts will not have vanished in a century. Long-distance television presents hard problems that are not likely to get easier. The waves do not bend around the earth, and cable or wave guides are expensive. Perhaps we should recall Thoreau's famous question on learning that Maine and Texas were to be joined by telegraph: "Who knows whether Maine and Texas have anything to communicate?" The fact that people can be brought together without actually physically congregating allows, as Thomson points out, of some degree of communal dispersion. The need for communication contributes to the growth of cities. But if political, commercial, administrative and other transactions could be conducted electromagnetically, a trend might set in toward decentralization.

How fast will our descendants travel? Faster than now but not as fast as some undoubtedly would like. The famous drag/

277

lift factor bedevils every effort to increase air speed; resistance due to the waves at the bow and the turbulence in the wake cuts into a ship's speed. Atlantic crossings in one and a half hours are foreseeable, as are train speeds of 100 to 150 miles per hour. While these by no means represent upper limits, it must be recognized that much higher speeds, especially for short distances, are simply not worth while. Increased speed means increased cost; the question is whether it is worth assuming the larger cost when there are end delays (at the start and finish of the journey) that are apt to account for half or more of the total elapsed time. Sir George does not foresee everyone hopping around in helicopters. At best they will not be as easy to operate as automobiles, and air space is not unlimited.

An interesting point arises in connection with ocean travel. A fast-moving ship produces waves that offer high resistance. This resistance increases rapidly and can be overcome only by lengthening the ship or by redesigning it so that it lifts itself out of the water and is partly an airplane. A better alternative is "to copy the fishes." They create practically no waves (and neither does a submarine). With nuclear power available, and with the elimination of surface "excrescences" that produce "skin friction," it should be possible to drive submarines at 70 or 80 knots with considerably less horsepower per ton than an Atlantic liner of the present day.

Sir George is pleasantly calm in discussing the prospects of space travel. In due time we shall get to the moon. The problems of getting there, landing, and returning are difficult but can be overcome. A very pretty idea is the possibility of making a fast jet propelled by particles emitted in nuclear fission. Fast electrons escaping from radioactive fission products, while themselves too light to form the material for an efficient rocket-jet, might be used to generate an electric field; this, in turn, could be used "to accelerate heavier charged bodies, either atoms or clusters of atoms, which provide the

278

actual material of the jet." The peacefully inclined will be interested to read that Thomson characterizes as "absurd" the emphasis Wernher von Braun, the imported German technician working for the U.S. Air Force, lays on the satellite station as an instrument of war. "I cannot see the least prospect of establishing a station that would not be destroyed almost at once by guided missiles from below, which would be far easier to construct than the station itself," says Sir George.

Interstellar travel "is not imminent but we may well be nearer to it in time than we are to Pekin man." The nearest star, Proxima Centauri, is 4.3 light years away. If velocities could be attained equal to half that of light, the journey would become feasible. Because of the relativistic contraction of time, the trip would be shorter from the point of view of the travelers than as measured on the earth. Roughly, 2.6 years might be saved. "This seems a small prize," observes Sir George, "for all the discomfort and risk" of a seventeen-year voyage. Presumably men will not undertake it just to stay young.

There is promise at last of doing something about the weather. In fact we may be able to change it before we can predict it. To illustrate the horrendous difficulties of forecasting, I quote Sir George's estimate that an electronic computer must perform 30 million individual operations to digest the data needed to calculate an hour stage of upper winds over a small section of the Atlantic. Meteorology is not only complex but filled with paradoxes as well. For example, the best opinion holds that an increase in the sun's radiation would not make the earth warmer but instead would lead to an ice age. The argument is that the earth's atmosphere is driven by solar radiation, that an increase of the latter would speed up the winds and so lead to more precipitation which, at the poles, would appear as snow.

Weather is "like a pencil balanced on its point." A slight tremor may make it fall; the problem is to get it to fall in the direction you want. One can approach the problem with brute force or delicately. With large amounts of energy available from atomic nuclei, one might attempt by flooding or by explosions to break up the Arctic ice or Greenland barrier and so change the climate over large areas. On the other hand an atom-thick layer of material laid on the earth would absorb enough solar radiation to affect the weather, perhaps profoundly. It may not be impossible to spread such a thin blanket over the whole earth (about a million tons of material would be required), but it is far from certain that the results would be desirable. The attempts that have been made, notably by Langmuir in the United States and Bowen in Australia, to produce artificial rainfall by seeding are impressive, but it is still too early, according to Sir George, to be confident of the usefulness of this method. The growing of plants in desert areas, thereby altering the absorption of radiation by the ground, has a high chance of modifying climate for the better. It is important to realize, however, that changes in the vegetation of one area may adversely affect the climate of an adjoining area, so that the method has political implications, international as well as domestic. In the world of today, and even more of tomorrow, nobody's back yard, it appears, is really his own.

There will be enough food to go around for quite a while, says Sir George. Chlorella has possibilities; so have yeasts. He also has much to say on other engrossing topics such as artificial mutations, domestication of animals (it is suggested that we train monkeys to pick our crops), control of population, and the prevention of old age. (Immortality, in his opinion, may not be desirable but it is not impossible.) On the social consequences of another century of technological progress Thomson's comments are inferior to the rest of his book. We might have been spared his astonishing views on "the future of the stupid" (he seems to favor a return to "domestic service" for those who cannot master the ways of the bright new world), on education and on the uses of leisure. They are at worst not very creditable and at best neither original nor very interesting. On one point, however, I agree: that (despite the efforts of the noted scholar-publicist, Dr. Rudolph Flesch) there is "far less real evidence on the best method [of teaching reading] than there is on the best sort of potato to grow."

The last chapter of the book, in which Sir George speculates on various aspects of the theory of communication, is fascinating. Very little is known about the working of the brain, but in time we shall know more. The big computers will help us, as will developments in electrophysiology, information theory and so on. Already enough is known to make one realize that the small, intricate fleshy lump in our skulls, "with its ten thousand million working parts and its countless possible interconnections, vastly exceeds anything we are ever likely to be able to make and is [utterly] unlike the unorganized masses we physicists study, which show at best the rather banal wallpaper patterns that crystals display." Suppose we attain a much deeper understanding of brain processes than at present; suppose we glimpse what is involved in the formation of ideas, habits, prejudices, desires and the like. Philosophers and even plain men have long esteemed self-knowledge as one of the greatest goods. Will they continue to esteem it if it should reveal the origins of altruism, tolerance, kindliness and other human virtues in terms of electrical circuits? Many things that men value may not survive such analysis. It is dangerous enough to understand the secrets of the atom, but suppose we understood — as we understand the operation of an electric washing machine — why we laugh or are patriotic or admire Matisse or embrace religion. How long could these values be maintained, once their genesis and nature were expressed as a circuit diagram? Sir George suggests that the possibility of reducing human responses to electromagnetic patterns need not necessarily impoverish life. We may learn new scales of values, learn to appreciate the intrinsic pro-

281

fundity of what seems to us trivial at the same time we perceive the trivial in what seems to us profound. "Consider the relation of Jove's thunderbolt to the fluttering of chaff round a piece of rubbed amber, consider how relatively easy we now find the movement of the planets and how hard it still is to understand the workings of a worm."

It cannot be said that we are remotely within sight of the relation between brain and what we call mind. That there is a simple one to one correlation between states of the brain and of consciousness has been assumed but never proved. Now doubt has been cast on this assumption by research on extrasensory perception, yielding evidence that Sir George regards as uncomfortably impressive. It may be necessary, if the evidence accumulates and stands up, drastically to revise our general scheme of thought. "The importance of the subject," says Sir George, "is enormous and much too little work is being done on it." Clearly, the frontiers of the mind lie much beyond us; not only is it possible that we use unsuspected modes of communication, but there are plain indications that "we are far from using our full mental potentialities." There is no permanent reason why only a few men should be bright and the rest mediocre, or worse. Why should the calculating wizard or the musical prodigy so often be an idiot in other respects? Is there no way to train special mental powers so that many will be able to do what can now be done only by few? "Do you not feel with genius that it alone thinks naturally, while the rest of us block our thoughts perversely with irrelevancies?" It is difficult to prophesy how men's brains will be improved, whether by drugs, by feeding in electrical impulses of the right kind, by new methods of education, by psychiatric stratagems employed in earliest youth, by selected mutations. But it will be done, Sir George believes, and since we have come a long way from being ammonites, in the head as well as in the body, we have proof that it can be done. That the brain which can foresee itself has a future that is not foreseeable is the greatest of the promises that lie ahead.

COMETS AND
THEIR ORIGINS

$\text{``}\mathbf{O}_\text{LD}$ MEN and comets," wrote Swift, "have been reverenced for the same reason; their long beards, and pretences to foretell future events." The Babylonians had a vague notion that comets moved like planets. Anaxagoras and Democritus attributed comets to "the combined splendour of a concourse of planets." Aristotle, a renowned chronicler of old wives' tales about astronomy and meteorology, maintained that comets were exhalations from the earth to the upper atmosphere. This hypothesis was so widely accepted that comets are not classified in Ptolemy's *Almagest* among the heavenly bodies.

But whatever the differences in the explanations of their physical nature, motions and causes, comets were until recent times universally regarded as presages: sometimes of happy

Figure d'yne Comette admirable yeue en l'air.

Comet of 1528, after Ambroise Paré. (Les Oeuvres d'Ambroise Paré, *Paris,*
1579, figure 341.) (From René Taton, Reason and Chance in Scientific Discovery, *by permission of Philosophical Library)*

augury, usually of death and disaster. The comet's sudden
and mysterious appearance, its flaming flight across the sky,
the swiftly changing aspect of its tail, its departure without a
trace all inspired awe and fed superstition. When a comet is
seen, "ther occurris haistily eftir it sum grit myscheif" (I
quote from the *Complaynt of Scotlande*, 1549); Shakespeare
(in *Henry VI*) wrote of "Comets imparting change of Times
and States"; Milton's image in *Paradise Lost* is perhaps the
most famous:

284

On the other side,
Incensed with indignation, Satan stood
Unterrified; and like a comet burned,
That fires the length of Ophiuchus huge
In the arctic sky, and from his horrid hair
Shakes pestilence and war.

The superstitions have all but vanished; myths about comets have been replaced by more fashionable irrationalisms. But the phenomenon itself is as impressive as ever, and as puzzling. The most recent edition of the *Britannica*, for example, reports that as late as 1946 — when the article on comets was written — no plausible explanation of comet formation had been proposed. R. A. Lyttleton, a Cambridge mathematician, now expounds a new, coherent and testable theory based on what he calls the "New Cosmology."* Sir Arthur Eddington, E. A. Milne, Henry Norris Russell, S. Chandrasekhar are among the leaders who contributed to the development of modern cosmological science; Fred Hoyle, Hermann Bondi, and Thomas Gold have been in the forefront of the work on the hypotheses underlying Lyttleton's study. I shall have more to say about this a little further on; but it may serve to a better understanding of the new theory if I first outline briefly Lyttleton's account of the dynamical and physical properties of comets. He has culled the voluminous literature on the subject and presents its main features in exemplary style.

Tycho Brahe was the first to demonstrate what had already been conjectured by that strange combination of genius and muttonhead, Jerome Cardan, namely, that comets are true celestial objects, far more distant than the moon. Tycho suggested that the path of the daylight comet of 1577 was a circle;

* R. A. Lyttleton, *The Comets and their Origin*, New York, 1953.

Kepler, not realizing that comets could return — and were therefore subject to the rules of planetary motion, which he himself had so brilliantly unfolded — concluded that they move in straight lines. The German astronomer Hevelius conjectured their path as parabolic; but it was Edmund Halley who, with the aid of Newton's theory of gravitation, finally solved the problem of cometary orbits. It will not do, of course, to overlook the ideas of the incomparable Robert Hooke. Rarely, if ever, did he pronounce on scientific subjects without saying something sensible. In this case he was extraordinarily prescient: comet tails, he supposed, are formed by the pressure of the sun's rays. This is the view now generally adopted by astronomers.

Comets move in highly complicated three-dimensional curves. When these are simplified for computational purposes into so-called two-dimensional osculating orbits, it is found that the orbits are conic sections: hyperbolas, parabolas, ellipses. The osculating orbit is the path a comet would follow if it were subject only to the simple inverse-square attraction of the sun. While the sun is the dominating influence and lies at the focus of the conic, other planets such as Jupiter cast the hem of their gravitational mantle over the comet, thus producing perturbations in its motion. In some instances the effect is very severe and results in a drastic change of orbit. Most osculatory orbits are parabolas, but a slight decrease or increase in the comet's speed converts the orbit into an ellipse or hyperbola, respectively.

Astronomers have now determined the orbits of about a thousand comets. The task is delicate and difficult. When a comet is newly discovered, a provisional path is calculated on the basis of its behavior for a few days. It is then kept under constant surveillance so that its position in space and that of its path with respect to the earth may be computed with increasing precision. A crucial reference datum in this computa-

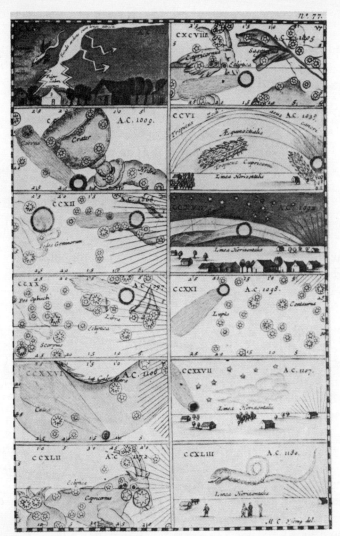

Imaginative pictures of some comets that appeared between 1000 and 1180, from the Theatrum Cometicum, *by Stanislav Lubienitz, Amstelodami, 1688, volume two, plate 77. Note particularly the fiery salamander (comet of the year 1000) and the fiery serpent (comet of 1180) which, in the popular imagination, accompanied the passage of these comets. (From René Taton,* Reason and Chance in Scientific Discovery, *by permission of Philosophical Library)*

tion is the comet's perihelion point: it is necessary to ascertain the exact time at which the comet passes or "will eventually pass" in closest proximity to the sun. For a comet to be periodic, its orbit must evidently be elliptical; on a hyperbolic or parabolic journey the traveler will come to us literally out of the nowhere and vanish forever into the beyond. Yet it is precisely this determination, dependent on minute differences of curvature and subject to various circumstances that limit the accuracy of observation, which is so difficult to settle. Careful tracking may lead to the conclusion that a cometary orbit is hyperbolic near the sun; however, when the orbit is "extended further outwards from the observable part and backwards in time, by calculations making due allowance for the influence of the planets (Jupiter is usually the main perturbing agency), in every case it has been found that the comet has in fact come in from a finite distance, and is therefore to be regarded as a reappearance of a permanent member of the solar system as far as its orbital motion is concerned." Since Kepler's third law — that the square of the time of revolution of a planet is proportional to the cube of its mean distance from the sun — describes (subject to planetary disturbances) the time of revolution of a comet in its orbit, it is easy to see that the period of a comet calculated from a necessarily limited arc is only weakly determined. This uncertainty helps to explain the extraordinary computational differences among leading astronomers. To cite only one illustration: the comet of 1680 has, according to the most accurate reckoning of the German astronomer Johann Franz Encke, a period not 170 years, as found by Euler, nor 575 years by Halley, nor 5,864 years by the Frenchman Alexandre Pingré, but in fact 8,814 years.

While we derive our knowledge of comets from those that have actually been seen, it must not be supposed that more than a small fraction of the comets chasing around the solar system ever have or ever will be seen. Short-period comets are defined as those traveling their circuit in less than 100 years;

even so, they may disappoint us after a number of visits, and fail to return. Brorsen's comet of 1846 (period 5.5 years) was not seen again after 1879, and Holmes's comet of 1892 (period 7 years) was not found in 1919 or 1928. Others, such as Encke's comet in 1944, are "missed through unfavorable circumstances but rediscovered at a later return." Halley's famous comet (period about 77 years) has been reasonably punctual for at least six centuries and may be looked for again in the spring of 1986. (A date not far from the beginning of George Orwell's dismal epoch.) Comets of moderate period — of which about forty are known — pay their homage to the sun every 100 to 1,000 years. But the great majority of comets have average orbital periods, according to the noted British expert A. C. D. Crommelin, of about 40,000 years. If a comet of this class flies into view, it is a safe prediction it will never be seen again; not only because of man's precarious tenancy of this planet, but because the computation based on the observation of a single sweep near the sun is too uncertain to justify a choice between a parabolic and an enormous elliptical orbit. It has been estimated that at least 300 long-period comets come to perihelion each century; if, then, the 40,000-year average is adopted, "we arrive at the amazing but inescapable conclusion that there must be at least 100,000 comets in the solar system with perihelion distances sufficiently small for them to become eventually observable." Moreover, there are many more with perihelion distances too great for the comets to be seen with present equipment if they remain in their existing paths.

About six or seven comets are discovered every year. The way to find one is to have the qualities that made a success of Phil the Fiddler — industry, zeal, attentiveness. Good equipment and luck also help. In 1896 the American astronomer Charles Perrine was at the Lick Observatory making loving observations of a comet he himself had discovered, when he received a telegram from Kiel stating the position of the comet

289

at that moment. But the telegram had been jumbled in transmission and gave an entirely wrong position, more than two degrees from the correct one. Perrine, not knowing the message had been twisted, pointed his telescope to the indicated place — and found a new comet. The devotion and perseverance of astronomers are typified by the almost incredible labors of Joseph de Lalande and his staff in computing the date of the return of Halley's comet in 1758. M. Lepaute, one of Lelande's assistants, tells the story:

"During six months we calculated from morning till night, sometimes even at meals; the consequence of which was that I contracted an illness which changed my constitution during the remainder of my life. The assistance rendered by Madame Lepaute was such that without her we never should have dared to undertake the enormous labor with which it was necessary to calculate the distance of each of the two planets, Jupiter and Saturn, from the comet, separately for every degree, for one hundred and fifty years."

The prediction was that the comet, having been delayed 100 days by the influence of Saturn and 518 days by that of Jupiter, would arrive at perihelion April 13, 1759. The actual date was only 32 days earlier, a tribute no less to the skill of the computers than to the theory.

What is a comet? Henry Norris Russell describes it as a "loose swarm of separate particles" accompanied by dust and gas. An observer sees it as a queer object with a head that is no head (and which, in any case, is sometimes missing), and a tail that conforms to no definition of a tail found in any dictionary (and may also be missing). The head, when it is present, consists of the coma, a faintly luminous cloud, which envelops a bright something called the nucleus. The coma is transparent and undergoes fantastic deformations as it passes the sun; the nucleus, thought to be made of "some kind of changing concentration of small particles," also can be expected to transform itself like a jinni — a fact that astronomers conveniently explain by classifying the nucleus as no more than an "apparent phenomenon." The most spectacular feature of a bright comet is its tail. This appendage has been known to stretch over 200,000,000 miles and to extend nearly 180° across the sky. Comets, like young women presented at St. James's, seem to put on their tails as they approach the sun. As a general rule the closer the perihelion distance, the more impressive the tail. As a comet approaches the sun, the tail streams behind, but beyond perihelion it precedes the comet. This curious behavior is explained on the theory that the pressure of the sun's radiation affects the particles of the tail in the way the wind affects a plume of smoke. A comet can have more than one tail: Chesaux's (1744) is said to have shown six, Borelly's comet (1903 IV) nine.

Comets are luminous partly because the small solid particles of which they are composed reflect, diffract and scatter the sun's light. But their spectrum, besides showing the familiar solar Fraunhofer lines, exhibits bright bands due to emission by molecules of energy originally absorbed from solar radiation. Photodissociation and other processes are involved in these complex effects.

There is considerable variation in the sizes and shapes of comets and in their masses. A few cometary giants are greater in volume than the sun itself. The more normal specimens range in diameter from 20,000 to 200,000 miles. In 1909 Halley's comet, which is not untypical, had an observed coma 14,000 miles across when distant three astronomical units* from the sun; at two units the coma had swelled to 220,000 miles; at perihelion it had shrunk to 120,000 miles; later, at one-unit distance, it had increased again to 320,000 miles. The prime example of shape-changing capacity is Biela's comet, a short-period wanderer (6.6 years). On its visit in 1846 — the same year Neptune was discovered — it first startled observers by making its entrance in pear-shaped form; 10 days

* The unit is the distance of the earth from the sun.

later it shattered the self-confidence of astronomers by dividing into separate comets, which continued to travel in practically the same orbit, one preceding the other by about 175,000 miles. The twins appeared again in 1852, now eight times farther apart. They have not been seen since. Biela's comet, by the way, provided another opportunity for the hapless Cambridge astronomer James Challis to demonstrate his exquisite incompetence. It was Challis who managed to miss finding Neptune after John Couch Adams told him where and when to look for it. His excuse was that he was searching for comets. To complete the tableau, he observed the twinning of Biela's comet but attributed it to an optical illusion and failed to publish his finding, explaining his oversight later on the ground he was too busy looking for Neptune.

I think it was Sir John Herschel who said that a comet could easily be packed in a portmanteau. Certainly the mass of a comet is insignificant compared to that of other celestial objects. The average comet, Lyttleton suggests, is only 1/10,000,000,-000th as massive as the earth, or, say, about 10^{18}g. (The earth's mass is 6×10^{27}g., Jupiter's 2×10^{30}g.) This comes to a million million tons, too large for a portmanteau, but too small to produce observable gravitational perturbations. Spread this mass of small stones or rocks through a volume equal to that of the sun, and there will be plenty of empty space between adjacent stones. On this assumption it is not surprising that comets are transparent. To be sure, the substance of the comet is not spread uniformly through it, and the mean-density figure 10^{-12}g. CM^{-3} (compared with 1.4 g. CM^{-3} for the sun) reflects neither the density of the head nor that of the tail. Lyttleton puts the density of the tail at "perhaps far less than 10^{-24}g. CM^{-3}" and supposes it to be made up of a mixture of dust and gas. Meteors and meteor rings are related to comets; one plausible hypothesis is that they are simply the "debris of disintegrated comets" whose material has become distributed along their orbits and now streams around the sun.

I come now to the new theory of cometary origins. It runs something like this. Genesis begins when the sun passes through a galactic dust cloud formed of material ejected in the explosion of a supernova. The sun's gravitational attraction starts the particles of the cloud, initially at rest, flying into hyperbolic orbits (i.e., hyperbolic relative to the sun). Fix your attention now on an imaginary line through the center of the sun parallel to the direction of its motion, the so-called accretion axis. The particles will converge, in their hyperbolic orbits, toward this axis and will collide when they are at and near the axis at points behind the sun. If we consider two inelastic particles of the same mass, moving from points on directly opposite sides of the accretion axis and "symmetrically placed with respect to the direction of motion of the sun," it can be shown not only that a head-on collision will occur on the axis, but that the results of the impact will be to nullify the transverse components of the particles' motion, to leave unaffected the radial components away from the sun along the axis, and "to reduce the originally hyperbolic energies of the particles to elliptic values, and thus bring about their capture by the sun." The multiplication of this and similar effects, involving hordes of colliding particles, creates a stream of material trailing behind the sun along the axis of its motion. Note, however, that part of the stream will flow *toward* the sun, part *away* from it. The reason is not difficult to grasp. Out to a certain distance from the sun, its gravitational pull will exceed the radial or escape velocity of the colliding particles and thus draw them inward along the accretion axis; beyond a certain "neutral point," material arriving at the axis will have a radial velocity sufficient to overcome the sun's captive power. These particles will eventually escape.

The accretion process forms a stream of reasonably high density, which is acted upon by two opposing forces (we are interested now only in the part of the stream flowing toward the sun). The sun's gravitational field pulls the stream one way; the internal gravitation of the material within the stream causes it to pull itself together lengthwise. We may assume, says Lyttleton, that irregularities of density in the dust cloud, and the general unstable nature of the accretion process, produce in the stream centers of attraction around which particles tend to cluster. While the stream's internal gravitation close to the sun is of negligible importance, farther out, with the sun's field diminishing as the inverse square of the distance, the self-gravitating force will promote not only the formation of local clusters but a lumping together and separation of the stream into segments. It is these segments, or aggregates, that, according to the theory, develop into comets. One asks why, since this entire part of the stream is flowing toward the sun, the comets are not sooner or later swallowed up by it. Lyttleton answers that much of the stream's material does fall into the sun, but some of the comets, held together by self-gravitation, escape this fate by the attraction of other planets, particularly Jupiter and Saturn. The effect of their gravitational fields is to endow nascent comets that are favorably placed with a sufficient angular momentum to sweep clear of the sun. He estimates that even if only a small percentage of all comets forming in the stream avoid extinction, "an average cloud might easily produce several thousand comets" that will survive.

The concluding portions of Lyttleton's book discuss the formation of comet tails (here, too, collisions are involved and the explanation of the process is drawn as a "simple consequence" of the theory), and compare the theory to earlier ideas of the origin of comets.

It is impossible not to admire the working out of the main system. It is coherent and — for the average reader, at any rate — excitingly persuasive. Astronomers and cosmologists, even those favorably disposed to the accretion hypothesis, will be harder to persuade; some, I am sure, will look at Lyttleton's conclusions with a fishy eye. None, however, will overlook the important fact that the central idea involves hypotheses "that can be subjected to quantitative tests — a feature hitherto completely absent from cometary theories." While these tests can only yield confirmation within orders of magnitude, their results are not to be despised; for as the British physicist, Sir Harold Jeffreys, once pointed out, most incorrect physical hypotheses fail order-of-magnitude tests "by many powers of 10."

This is not always an easy book, but I recommend it highly. It impressed me so much that I now appreciate the feelings of a certain enthusiastic young lady from New Jersey (reported by Mary Proctor in her book on comets) who, on the appearance of Halley's comet in 1910, declared her intention of following it "wheresoever it went." Only the counsel of timid friends and "temporary seclusion" in an asylum deterred her from this gallant pursuit.

Index · Volume Two

NOTE: Small capitals are used for subjects to which an entire section is devoted. Page numbers in italics refer to such sections, roman to other mention.

Civilizations: "arrested," 185; *see also* Society

Clairvoyance: *see* Extrasensory perception

Clifford, William Kingdon: on causality, 127, 141; on science, 126

Climate: in the future, 235; modification of, 280

Cold war: 193

Coleridge, Samuel Taylor: 30, 143; *Rime of the Ancient Mariner, The,* 254, 268-269

Collectivism: 35

Columbia University: atomic inspection study at, 198-203; atomic research at, 206

COMETS: *283-295;* illustrations of, 284, 287; orbits of, 286, 288; origins of, 285, 293-295; periods of, 288-290; sizes of, 291-292; superstitions about, 283-285, 287; tails of, 286, 290

COMETS AND THEIR ORIGIN, THE (Lyttleton): reviewed, *283-295*

Common sense: science and, 126

Communication: future of, 277; theory of, 281-282

Communism (Marxism): 70; law under, 165-172; Russell on, 46, 49; in U. S., 179-181

Complementarity: 122, 126

Compte, Auguste: 31

Compton effect: 138

Computers: weather, 279; *see also* Calculating machines

Conant, James B.: 214, 215

Conic curves: 7-8

Conrad, Joseph: Russell on, 50

Consciousness: electrical patterns and, 250-251; evolution and, 92-97; habit and, 90-91; physical basis of, 88-91, 282

Conservation of mass-energy: 272

Cook, (Capt.) James: 263

Cornelius Nepo: 26

Cortical rhythm: 250-251

Coulomb, Charles Augustin de: 132

Cowper, William: 26

Craik, Kenneth: on thinking, 252

Creativity: *see* Scientific discovery

Croce, Benedetto: philosophy of, 72-73

Crommelin, A. C. D.: on periods of comets, 289

Crystal lattice: 118-119

Cycloid: 19

D'Aiguillon, Duchesse: 15

Dalibray, Charles: 7

Darwin, Charles: 182; on albatrosses, 263; on natural selection, 90

Darwin, Charles Galton: THE NEXT MILLION YEARS (reviewed), *230-239;* 270

Darwin, (Sir) George: 42

DDT: 176

Dean, Gordon: on misconception that atomic energy exclusively American, 217

De Broglie, Louis: 136, 140, 153; on causality in physics, 113-114, 122

De Fermat, Pierre: 19; on probability, 17, 241

Defoe, Daniel: *Robinson Crusoe,* 25

De Gusmão, Lourenço: 160

De Lalande, Joseph: 147; computes return of Halley's comet, 290

De Laplace: *see* Laplace

De la Rue, Auguste: 158

De Méré: *see* Méré

Democritus: 115, 283

De Montaigne: *see* Montaigne

De Roannez, Duc: 14, 16

De Roberval, Personne: 7

Desargues, Gérard: Pascal and, 7-8

De Saron, Bochart: 147

Descartes, René: 15, 56, 70; on the body, 90-91; *Discourse on Method,* 19; on the vacuum, 11, 13

"Determined" and "undetermined" descriptions in physics: 133-135

Determinism: 75, 107; Newton's laws as, 114; precognition as, 243; *see also* Causality, Uncertainty principle

DETERMINISM AND INDETERMINISM IN MODERN PHYSICS (Cassirer): 122; reviewed, *127-141*

Deterrence: by genocidal weapons, 202

De Tocqueville, Alexis: 171; *Democracy in America,* 36

301

303